AF847650

Fluid Mechanics and Its Applications

Founding Editor

René Moreau

Volume 141

Series Editor

André Thess, German Aerospace Center, Institute of Engineering Thermodynamics, Stuttgart, Germany

The purpose of this series is to focus on subjects in which fluid mechanics plays a fundamental role. As well as the more traditional applications of aeronautics, hydraulics, heat and mass transfer etc., books will be published dealing with topics, which are currently in a state of rapid development, such as turbulence, suspensions and multiphase fluids, super and hypersonic flows and numerical modelling techniques. It is a widely held view that it is the interdisciplinary subjects that will receive intense scientific attention, bringing them to the forefront of technological advancement. Fluids have the ability to transport matter and its properties as well as transmit force, therefore fluid mechanics is a subject that is particulary open to cross fertilisation with other sciences and disciplines of engineering. The subject of fluid mechanics will be highly relevant in such domains as chemical, metallurgical, biological and ecological engineering. This series is particularly open to such new multidisciplinary domains. The median level of presentation is the first year graduate student. Some texts are monographs defining the current state of a field; others are accessible to final year undergraduates; but essentially the emphasis is on readability and clarity.

Springer and Professor Thess welcome book ideas from authors. Potential authors who wish to submit a book proposal should contact Dr. Mayra Castro, Senior Editor, Springer Heidelberg, e-mail: mayra.castro@springer.com

Indexed by SCOPUS, EBSCO Discovery Service, OCLC, ProQuest Summon, Google Scholar and SpringerLink

Mathias Bechert · Benoit Scheid

Draw Resonance

Springer

Mathias Bechert
Institute of Polymer Materials
Friedrich-Alexander-Universität
Erlangen-Nürnberg
Erlangen, Germany

Benoit Scheid
Université libre de Bruxelles
Brussels, Belgium

ISSN 0926-5112 ISSN 2215-0056 (electronic)
Fluid Mechanics and Its Applications
ISBN 978-3-031-85567-2 ISBN 978-3-031-85568-9 (eBook)
https://doi.org/10.1007/978-3-031-85568-9

Partly published previously as: Influence of Process and Material Parameters on the Draw Resonance Instability, Mathias Bechert, FAU University Press, 2017, Language: English, ISBN 978-3-96147-067-9 also available online on https://opus4.kobv.de/opus4-fau/frontdoor/index/index/docId/8923

© Springer Nature Switzerland AG 2025, corrected publication 2025

This work is subject to copyright. All rights are solely and exclusively licensed by the Publisher, whether the whole or part of the material is concerned, specifically the rights of translation, reprinting, reuse of illustrations, recitation, broadcasting, reproduction on microfilms or in any other physical way, and transmission or information storage and retrieval, electronic adaptation, computer software, or by similar or dissimilar methodology now known or hereafter developed.
The use of general descriptive names, registered names, trademarks, service marks, etc. in this publication does not imply, even in the absence of a specific statement, that such names are exempt from the relevant protective laws and regulations and therefore free for general use.
The publisher, the authors and the editors are safe to assume that the advice and information in this book are believed to be true and accurate at the date of publication. Neither the publisher nor the authors or the editors give a warranty, expressed or implied, with respect to the material contained herein or for any errors or omissions that may have been made. The publisher remains neutral with regard to jurisdictional claims in published maps and institutional affiliations.

This Springer imprint is published by the registered company Springer Nature Switzerland AG
The registered company address is: Gewerbestrasse 11, 6330 Cham, Switzerland

If disposing of this product, please recycle the paper.

Instability is necessary to progress.
If we stay put, we go backwards.

—Coluche

To my mother,
And to Soline, Zephyr, Hippolyte and Ulysse,
My dearest wife and children.

—Benoit Scheid

To those who tirelessly seek, collect, and share knowledge.

—Mathias Bechert

Foreword

Optical fibers underpin the entire modern communication system. Their robust and fast fabrication is however jeopardized by what is commonly called an instability, disturbances that spontaneously grow from infinitesimal, minute, initial perturbations. Exponentially growing instabilities are quite common in hydrodynamics where they often originate in the transition to turbulence as well as in several free-interface flows, like capillary jets destabilized by the Rayleigh-Plateau instability, thin coating films suspended on the underside of a substrate dripping as the consequence of the Rayleigh-Taylor instability, or falling films running down an incline forming roll waves originating from the Kapitza instability.

Somewhat surprisingly, the draw instability is not included in classical textbooks on hydrodynamic instabilities, despite the extreme elegance of its first rationalization in the 1970s. How can such an instability, which combines a particularly important application in manufacturing, a rather simple mathematical explanation of its essence as well as a formidable testbed for the incorporation of several important additional physical effects, have been neglected for so long? This book repairs this injustice!

Having contributed to several refinements of the minimal physical mode over the last decade, Benoit Scheid and Mathias Bechert are uniquely placed to take the reader on board along this route of increasing complexity. Starting from the self-explanatory set of equations, they include physical effects one by one: gravity, inertia, temperature, geometry, viscoelastic, and capillary effects are progressively analyzed, introducing the necessary dimensionless numbers enabling their respective weighing and seizing.

The authors have made a real effort for a systematic and unified framework to fiber drawing and film casting, ensuring consistent notations throughout the chapters. One of the challenges was to manage fibers and films in parallel, which share as many similarities as differences: while the apparent geometrical simplicity of films who can be modeled at first sight in two dimensions is jeopardized by transverse effects, axisymmetric fibers keep their symmetric geometry but are affected by capillary forces.

Maybe because it is written by the joined hands of a scholar and of an applied engineer, this book is translational in essence: each dimensionless physical insight is translated into curves of interest to the practitioners. The authors excel in the simultaneous focus on the physical analysis and on the obtention of guidelines and rules for the engineer facing a practical question at hand. This monograph is therefore both of fundamental interest and practical use. This is quite of a unique combination.

Given the significance of draw resonance instability as a fundamental problem in fluid dynamics and its critical role in the glass and polymer industries, this monograph fills an important literature gap with style and conciseness. I encourage the reader to dwell into this little (but complete!) story on draw instability, whether for the sake of exploring an astonishing fluid phenomenon or for better understanding of the origins of important fabrication constraints.

Lausanne, Switzerland
September 2024

François Gallaire

The original version of the book has been revised. A correction to this book can be found at https://doi.org/10.1007/978-3-031-85568-9_13

Preface

This book would have not been possible without the initial impulse of Saint-Gobain Recherches in collaboration with Howard Stone with whom Benoit Scheid did his postdoc from 2007 to 2009 at Harvard University. The project granted by the company aimed at investigating possible thermal instabilities preventing to make flat glass below 1 mm thick using the float glass process. The float glass process consists of stretching a cooling sheet of molten glass floating on molten tin such as to decrease the thickness of the sheet before being frozen at the glass transition. Stretching viscous sheets are subject to an instability called draw resonance occurring when the stretching velocity overpass a given threshold, which exponentially depends on the temperature, as the viscosity does. Draw resonance makes the viscous sheet oscillates around its centerline with a wave propagating periodically from the inlet to the outlet. Demonstrating that the heat transfer could non-monotically stabilize the draw resonance was our first contribution to that problem.

The major impulse that really put us on the track of this book arose then in the context of the Ph.D. thesis of Mathias Bechert from 2012 to 2017 at the Friedrich-Alexander-University Erlangen-Nuremberg in the Institute of Polymer Materials chaired by Dirk Schubert. Together we addressed one by one all effects that can have an influence on the draw resonance in fiber drawing and film casting, namely, gravity, inertia, surface tension, lateral aspect ratio, and viscoelasticity. A final impulse for the book came during the collaboration with 3B-Fiberglass company who granted the postdoc of Julien Philippi from 2018 to 2020 at the Université Libre de Bruxelles and with whom we addressed thermal effects for the fiber drawing as it is crucial to understand the breakage events that impede the production of glass fibers for decades.

Given the significance of draw resonance instability as a fundamental problem in fluid dynamics and its critical role in the glass and polymer industries, we have chosen to compile all our work on this topic, spanning over more than a decade of research, into this monograph. Our goal is to provide a comprehensive resource with a unified approach to fiber drawing and film casting, ensuring consistent notation throughout the chapters. This book is intended for undergraduate and graduate students in Physics, Applied Mathematics, and Engineering, as well as researchers seeking to deepen their understanding of draw resonance, a topic that has captivated us and will, we hope, prove equally intriguing to others.

Erlangen, Germany
Brussels, Belgium
September 2024

Mathias Bechert
Benoit Scheid

Acknowledgements This Monograph originates from several collaborations and the authors would like to acknowledge the following.

Sarah Quiligotti, Binh Tran and René Gy from Saint-Gobain Research for their deep insights into the float glass process; Howard Stone from Harvard University for his inspiring guidance in the universe of fluid physics and its intrication with applied mathematics; Dirk Schubert from Friedrich-Alexander-University Erlangen-Nuremberg for his support; Manuel Alves and Rob Poole for fruitful discussions on viscoelastic effects; Quentin Chouffart, Christophe Waucquez and Dimitri Laurent from 3B-Fiberglass for their deep insights into glass fiber drawing; Julien Philippi for his contribution to glass fiber drawing with thermal effects; François Gallaire for prospective discussions on weakly nonlinear analysis.

Part of the research reported in this monograph was sponsored by Saint-Gobain Research and 3B-Fiberglass companies. Benoit Scheid is FNRS Research Director.

Competing Interests The authors have no competing interests to declare that are relevant to the content of this manuscript.

Contents

1 Introduction 1
1.1 Fiber Drawing and Film Casting Processes 1
1.2 Draw Resonance 5
1.3 Literature Review 7
1.3.1 Theory 8
1.3.2 Experiments 9
1.3.3 Reviews 11
1.4 Observing the Critical Draw Ratio and Frequency 12
1.5 Modeling Draw Resonance and Outline 15

2 One-Dimensional Newtonian Models 19
2.1 Fiber Model 21
2.2 Film Model 25

3 Linear Stability 29
3.1 Pure Viscous Case 30
3.2 Linear Stability: General Framework 35
3.2.1 Chebyshev Collocation Method 37
3.2.2 Continuation Method 38
3.2.3 Additional Remarks on the Linear Stability Analysis 40
3.3 Physical Mechanism and Alternative Stability Criteria 40
3.3.1 Kim's and Jung's Criteria 41
3.3.2 Unity-Throughput Time t_L 45
3.3.3 Time Delay Δt_h and General Phase Shifts 46

4 Influence of Gravity and Inertia 47
4.1 Influence of Gravity on Draw Resonance 48
4.2 Influence of Inertia on Draw Resonance 51

5 Role of Surface Tension in Fiber Drawing 59
5.1 Influence of Surface Tension on Draw Resonance 59
5.2 Destabilization Mechanism of Surface Tension 61

6 Finite Width Effects in Film Casting ... 65
6.1 Geometrical Effects ... 65
6.2 Finite Width Film Model ... 66
6.3 Influence of Neck-in on Draw Resonance ... 70

7 One-dimensional Viscoelastic Models ... 77
7.1 Viscoelastic Constitutive Equations ... 77
7.2 Literature Review ... 80
7.3 Viscoelastic Film Models ... 82
7.3.1 Model Equations ... 82
7.3.2 Initial Stress Boundary Condition ... 84

8 Influence of Viscoelasticity ... 87
8.1 Linear Stability Analysis ... 87
8.1.1 UCM Model ... 87
8.1.2 Giesekus Model ... 89
8.1.3 PTT Model ... 94
8.1.4 Comparison and Evaluation of Constitutive Equations ... 99
8.2 Effective Elongational Viscosity ... 100
8.3 Generalized Newtonian Fluid Models ... 103
8.3.1 The Power-Law Model ... 103
8.3.2 The Carreau–Yasuda Model ... 105
8.3.3 Influence of Strain Hardening/Thinning ... 110
8.3.4 Elastic Effects and Stability Mechanism ... 113
8.4 Conclusions on the Choice of the Constitutive Equation ... 114

9 Practical Scaling and Stability Maps ... 115
9.1 Practical Control Parameters ... 115
9.2 Infinite-Width Film Casting ... 119
9.2.1 Newtonian Model ... 119
9.2.2 Viscoelastic Model ... 124
9.3 Finite Width Film Casting ... 128
9.3.1 Combined Effects of Inertia, Gravity and Neck-in ... 128
9.3.2 Viscous–Inertia–Neck-in Regime ... 131
9.3.3 Viscous–Gravity–Neck-in Regime ... 134
9.4 Fiber Drawing ... 135
9.4.1 Newtonian Model ... 136

9.4.2 Viscoelastic Model 141
9.5 Choosing the Appropriate Stability Map for a Particular Process 145

10 Non-isothermal Film Casting 147
10.1 Heat Transfer in Draw Resonance 147
10.2 Non-isothermal Model 148
10.2.1 Flow Field 148
10.2.2 Temperature Field 150
10.2.3 Depth-Averaged Viscosity 152
10.2.4 Far-Field Ambient Temperature 153
10.3 Stability Problem 154
10.3.1 Base State 154
10.3.2 Perturbation Equations 155
10.4 Neutral Stability 155
10.5 Prescribed Temperature Limit 159
10.6 Mechanism of Draw Resonance in the Non-isothermal Case 160
10.7 Stability Criteria in Non-isothermal Conditions 162
10.8 Radiative Effect 164

11 Non-isothermal Fiber Drawing 165
11.1 Non-isothermal Model 165
11.2 Stability Analysis 171
11.3 Comparison with Film Casting in the Pure Viscous Case ($F^* = 0$) with $h = c^{st}$ 173
11.4 Purely Viscous Non-isothermal Model ($F^* = 0$) 174
11.4.1 Influence of the Ambient Temperature 175
11.4.2 Influence of Non-constant Heat Transfer Coefficient 175
11.5 Full Non-isothermal Model ($F^* > 0$) 177
11.5.1 Preliminary Study for the Full Model: Influence of the Fluidity F^* 177
11.5.2 Influence of the Aspect Ratio Λ 178
11.5.3 Influence of the Inlet Temperature θ_0 and Stiffness Parameter k 179
11.5.4 Influence of the Radiative Heat Transfer 181
11.6 Few Perspectives 182

12 Conclusions and Outlook 185

Correction to: Draw Resonance C1

Appendix A: Eigenvalue Spectrum Determination 189

Appendix B: Continuation Method Using AUTO-07P: Tutorial 191

References 193

Acronyms and Symbols

Acronym

CY	Carreau–Yasuda
FFT	Fast Fourier Transformation
GNF	Generalized Newtonian Fluid
LCB	Long Chain Branched
PE	Polyethylene
PET	Polyethylene terephthalate
PP	Polypropylene
PP1	Polypropylene Moplen HP556E
PP2	Polypropylene Moplen HP501L
PTT	Phan-Thien–Tanner
UCM	Upper-Convected Maxwell
WM	White–Metzner
XPP	EXtended Pom-Pom

Symbol

a	Fiber cross-sectional area
A	Fiber cross-sectional area perturbation, film aspect ratio
b	Parameter for the ambient temperature
$\mathbf{A}$	$3(J+1) \times 3(J+1)$ matrix of Eq. (3.27)
b_0, b_1, b_2	Fit parameters of Eq. (9.15)
$\mathbf{b}$	Body force vector
B	Fit parameter of Eqs. (4.4), (4.8), and (9.14), constant in the Vogel-Fulcher-Tamman law

$\mathbf{B}$	$3(J+1) \times 3(J+1)$ matrix of Eq. (3.27)
Bi	Biot number
$c, \tilde{c}, \tilde{c}_f, c_{\mathrm{L}}, c_{\mathrm{G}}, c_{\mathrm{CY}}$ $c_{\mathrm{PTT}}, \tilde{c}_{\mathrm{PTT}}, \tilde{c}_{\mathrm{PTT},f}$ c_1, c_2, c_3	Auxiliary variables
c (index)	At criticality
c_p	Heat capacity
C	Fit parameter of Eqs. (4.4), (4.8), and (9.14)
Ca, Ca^*	Capillary number and rescaled capillary number as defined by Eq. (2.6)
D	Fit parameter of Eq. (5.3)
$\mathbf{D}$	Rate-of-deformation tensor as defined by Eq. (1.8), derivation matrix for Chebyshev coefficients
De	Deborah number as defined by Eq. (7.9)
Dr	Draw ratio
Dr_{c}	Critical draw ratio
Dr^{eff}	Effective draw ratio
$\mathbf{e}_x, \mathbf{e}_y, \mathbf{e}_z, \mathbf{e}_r, \mathbf{e}_\varphi$	Unity vectors
E	Fit parameter of Eq. (5.3)
E_A	Activation energy
f	Dimensionless force or tension, oscillation frequency
f^*	Film tension
f_0, f_1	Fit parameters of Eq. (9.15)
F	Dimensionless fluidity as defined by Eq. (9.3), force perturbation
F_{us}	Threshold value of F to unconditional stability
Fr	Froude number as defined by Eq. (2.6)
g	Gravitational acceleration
G'	Storage modulus
G''	Loss modulus
h	Film thickness, heat transfer coefficient in Chap. 10
h_0	Inlet film thickness
H	Film thickness perturbation
$\mathbf{I}$	Identity matrix
I (index)	Imaginary part of a complex quantity
J	Order of discretization of the eigenvalue problem
J_e^0	Linear steady state recoverable compliance
k	Heat conductivity, stiffness parameter of the ambient temperature θ_h, CY parameter of Eq. (8.18), dimensionless activation energy
k_{a}	Heat conductivity of air
ℓ	Half-film width, nozzle height

ℓ_0	Inlet half-film width
ℓ_c	Capillary length
L	Casting length, half film width perturbation
m	CY parameter of Eq. (8.18), extrudate mass
$\mathbf{m}$	Fluid surface normal vector
$\mathbf{M}$	Chebyshev transformation matrix
M_n	Number average molecular weight
M_w	Weight average molecular weight
n	Power-law index
$\mathbf{n}$	Fluid surface normal vector
N	Normal stress difference perturbation
p	Isotropic pressure, critical exponent of Eq. (8.8)
p_0, p_1, p_2	Fit parameters of Eq. (8.22)
P	Draw resonance oscillation period
Pe	Péclet number
Pr	Prandtl number
q	Flow rate, fit parameter of Eq. (8.8)
Q	Flow rate perturbation, dimensionless inlet velocity as defined by Eq. (9.8)
r	Radial variable
R	Fiber radius, gas constant
R (index)	Real part of a complex quantity
R_0	Inlet fiber radius
Re	Reynolds number as defined by Eq. (2.6)
Re_{us}	Threshold value of Re to unconditional stability
R_{adj}	Adjusted correlation coefficient
s (index)	Steady state
S	Dimensionless surface tension parameter as defined by Eq. (9.6)
St	Stanton number
t	Time
t_L	Unity-throughput wave traveling time
t_s	Steady state xx-component of the extra stress tensor
T	Draw resonance oscillation period, xx-component of the extra stress tensor perturbation, temperature
T_0	Inlet temperature
T_{a}	Ambient temperature
T_i	i-th Chebyshev polynomial of first kind
Tr	Trouton ratio
u	Axial velocity
u_0	Inlet axial velocity
U	Axial velocity perturbation

$\mathbf{v}$	Flow velocity field
v	Velocity component in film thickness direction, radial velocity component
V	Extruder screw revolution speed
w	Velocity component in film width direction
x	Axial variable
$\hat{x}_i$	i-th Chebyshev collocation point
y	Film thickness direction variable
z	Film width direction variable
$\mathbf{z}$	$3(J+1)$-dimensional state vector of Eq. (3.27)
α	Fiber parameter, nonlinear Giesekus parameter, heat transfer coefficient in Chap. 11
β	Film parameter, inverse growth rate of Eq. (1.3)
γ	Surface tension
Γ	Viscosity perturbation
δ	Fit residual
ε	Nonlinear PTT parameter, emissivity
ε_0	Typical average emissivity
ζ	Neck-in parameter
η	Shear viscosity
η_0	Newtonian shear viscosity, inlet viscosity
η_∞	Constant in the Vogel-Fulcher-Tamman law
η^*	Complex shear viscosity
$\hat{\eta}$	Depth-average viscosity
θ_L	Maximum- or minimum-thickness wave traveling time
θ	Depth- or cross-section-averaged temperature
θ_c	Constant in the Vogel-Fulcher-Tamman law
θ_0	Inlet cross-section-averaged temperature
θ_h	Reference temperature for convective heat transfer
θ_{h_∞}	Dimensionless far-field temperature in convective heat transfer
θ_ε	Reference temperature for radiative heat transfer
$\theta_{\varepsilon_\infty}$	Dimensionless far-field temperature in radiative heat transfer
Θ	One-dimensional temperature perturbation
κ	External field of Eq. (1.3)
λ	Relaxation time
λ_{CY}	Characteristic time of the CY function (8.18)
λ_{PL}	Characteristic time of the power-law (8.14)
Λ	Dimensionless relaxation time as defined by Eq. (9.1), fiber aspect ratio (L/R_0)
μ	Elongational viscosity, temperature-sensitive parameter of the viscosity

μ^{eff}	Effective elongational viscosity
ν	Normal stress difference
ν_{a}	Kinematic viscosity of air
ξ	Nonlinear Larson parameter, fit error
Ξ	Reduced mean square fiber diameter fluctuation as defined by Eq. (1.2)
ρ	Material density
σ	Stefan-Boltzmann constant
$\boldsymbol{\sigma}$	Stress tensor
$\boldsymbol{\tau}$	Extra stress tensor
τ_L	Fluid residence time
φ	Angular variable
ω	Complex eigenvalue

Chapter 1
Introduction

Abstract This chapter introduces the fiber drawing and film casting processes where the phenomenon of draw resonance can be observed. The physics of draw resonance is next described, followed by a review on dedicated experiments and more insights on the way critical draw ratio and frequency at threshold can be experimentally determined. The chapter ends by introducing the governing equations and the important modeling parameters influencing the draw resonance, which brings us to the outline of the book.

1.1 Fiber Drawing and Film Casting Processes

Most of the questions treated in fundamental research originate from actual phenomena in nature or man-made processes. While this may often be disguised due to strong abstraction or simplification, it is generally reasonable to keep the origin in mind to prevent research as an end in itself. The instability phenomenon treated in this book, which is commonly known as *draw resonance*, has its origin in basically three industrial processes: fiber drawing,[1] film casting and film blowing. Within all three processes, molten material is stretched to obtain products of small dimensions, i.e., fibers of small diameter and films of small thickness. While fiber drawing and film casting are applied for both glass and polymer materials and exhibit large similarities in their basic ingredients, film blowing is somewhat special as it is exclusively used for polymer materials, and as the fundamental setup differs more from the former two. For this reason we will concentrate on fiber drawing and film casting in this book.

Figure 1.1 sketches a typical industrial plant used to produce polymer yarns. Granular material is homogeneously molten in an extruder and afterwards transported by a melt line to a melt pump. This device reduces fluctuations and ensures a constant throughput of the material, which is then extruded through the *spinneret* consisting of one or several nozzles, often referred to as *dies*, which are mostly of circular

[1] We use the more general terminology "fiber drawing" instead of "fiber spinning" throughout this book, which will be explained in the following section.

© Springer Nature Switzerland AG 2025
M. Bechert and B. Scheid, *Draw Resonance*, Fluid Mechanics and Its Applications 141,
https://doi.org/10.1007/978-3-031-85568-9_1

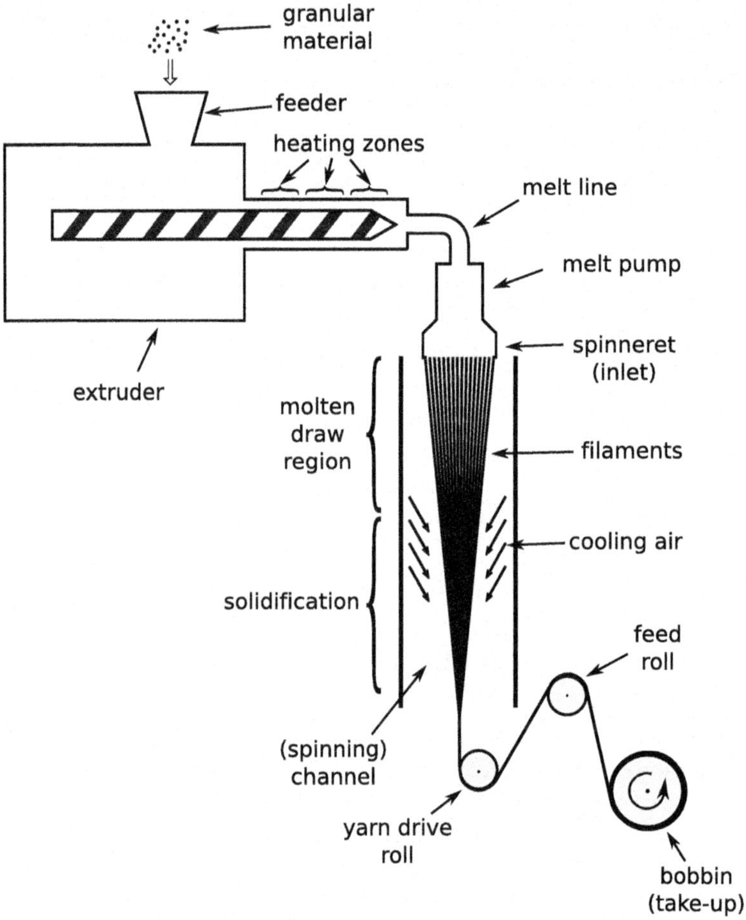

Fig. 1.1 Sketch of a fiber spinning setup, one example of fiber drawing processes used to produce polymer yarns. The polymer melt is extruded through a spinneret consisting of several dies and afterwards stretched by the bobbin, which rotates at high speed. During the stretching, the filaments are solidified by cooling air in the channel

shape. The so-called filaments are afterwards stretched and cooled by blowing air in a channel, which can have a length of several meters. The stretching is hereby driven by the bobbin, a roll rotating at high-speed and winding up the filaments after they are guided by yarn drive and feed rolls. In practical application, the take-up speed is very large so that the fiber diameter can be reduced by several orders of magnitude. The cooling air serves primarily as a stabilizer of the process and leads to a solidification of the material within the channel. Consequently, most of the filament stretching takes place in the so-called molten draw region.

Due to the rotating bobbin, this way of producing fibers is classically called *fiber spinning*. It has to be noted, however, that there exist a multitude of different setups to produce fibers, depending on the actual application purpose. Variations may occur, for instance, in the number of dies, reaching from single *monofilaments* to several hundreds or thousands *multifilaments*. Moreover, the *drawing length*, which is defined as the length between the spinneret and the yarn drive roll, can be very short, in the range of few centimeters, as well. In this case the channel is omitted and the yarn drive roll is often placed in a water bath in order to solidify the fiber. Finally, it is also possible to stretch the fiber by blowing air instead of a bobbin. This becomes especially relevant for the production of so-called non-woven materials, which consist of sheets of randomly arranged and afterwards calendered fibers. We will use the general term *fiber drawing* to describe this entire family of processes, which includes fiber spinning as a particular case. In literature, the wording is often rather imprecise and fiber spinning is mostly used for all fiber producing processes.[2]

In the glass fibers manufacturing, the process is comparable to the one presented in Fig. 1.1 with the difference that the glass furnace is at the top such as molten glass flows through a matrix of nozzles (the spinneret) by hydrostatic pressure. The glass filaments are cooled down by air convection until they solidify and can then be pulled down by a winder in order to considerably reduce the cross-section area of the resulting fibers.

Let us now have a look at the film casting process. A typical example of a polymer film casting setup is depicted in Fig. 1.2. Analogue to fiber drawing, molten material made of granulate is extruded through a die, but this time a slit die with mostly rectangular cross-section. The thickness of the die is very small, in the range of millimeters, while the width can be up to the order of few meters. After a gap or *casting length*, which spans often few decades of centimeters, the film is taken-up by one or several so-called *chill rolls*, which also serve as cooling devices. Similar to the bobbin in fiber spinning, the chill roll rotates at high speed in order to reduce the film thickness by stretching. Apart from the film thickness, the film *width* usually shrinks from the inlet to the take-up due to the strong elongational deformation caused by the pulling, which is typically referred to as *neck-in* or *necking*. This necking phenomenon makes the sheet thickness thicker than expected from a pure parallel stretching. To prevent this response, it is common in processes such as polymer film casting to keep the distance between the inlet and the take-up roll as short as possible, so that the aspect ratio between the width and the length is usually of order of unity in such processes. Like in fiber drawing, various forms of film casting, depending on the final product and the material, can be found, including vertical and horizontal casting directions.

The float glass process for instance is an horizontal film casting process (top view). As sketched in Fig. 1.3, molten glass is poured on a bath of molten tin. Because glass is lighter than tin, it floats, hence the name *float glass*, the tin bath simply serving as an hydrostatic support. Indeed, molten tin being at least four orders of magnitude less viscous than molten glass, no shear exists between both liquids and the flow in the glass film remains purely extensional.[3] As it flows, the glass cools down to the glass transition temperature at which it becomes solid. The resulting *glass ribbon*

---

[2] Note that in all fiber spinning processes, the fiber never spins. What spins are the rolls at take-up.

[3] An extensional flow is characterized by a uniform velocity profile across the flow direction, also referred to as plug flow.

can then be pulled out of the bath and edge rollers are used to prevent the shrinking of the film until it solidifies (Pilkington 1969). In practice, the rollers are placed closer

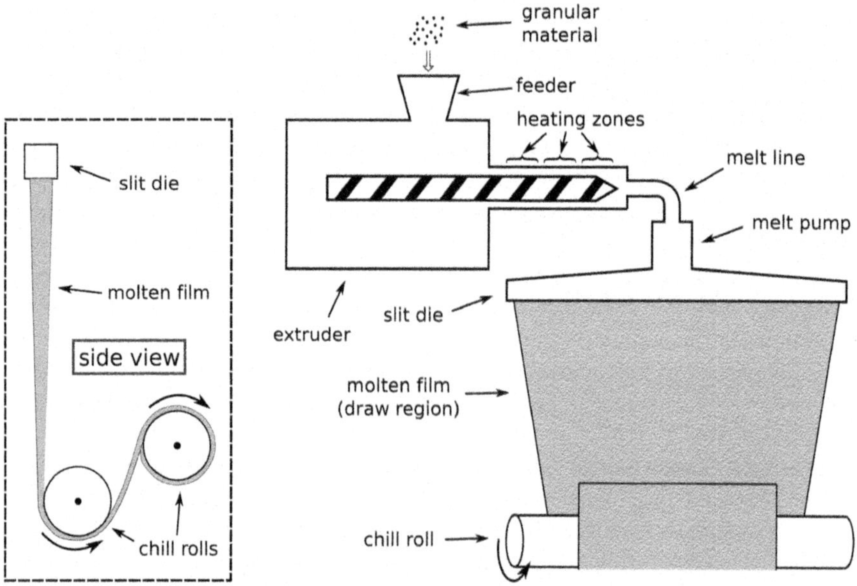

Fig. 1.2 Sketch of a film casting setup used to produce polymer sheets. The polymer melt is extruded through a slit die and afterwards stretched by one or several chill rolls, which rotate at high speed and cool down the material

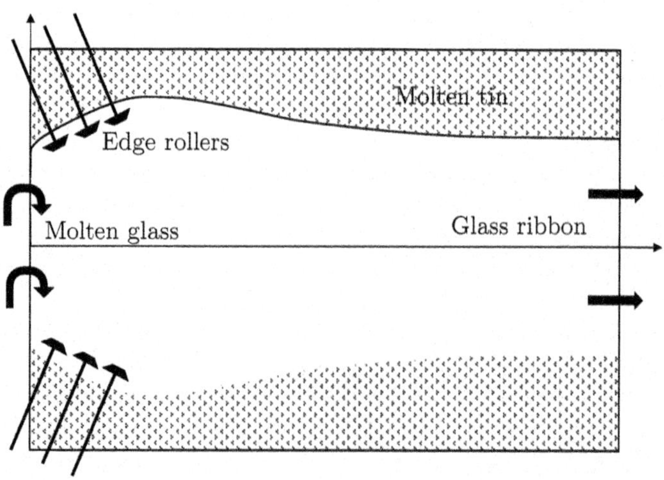

Fig. 1.3 Sketch of the float glass process (top view). Molten glass is poured on a bath of molten tin and forms a glass sheet that solidifies into a glass ribbon, which can thus be pulled out of the bath. Edge rollers are used to prevent shrinking (or *necking*) of the glass sheet

to the inlet than to the take-up. Therefore the glass film still shrinks in width between the last edge roll and the take-up.

1.2 Draw Resonance

We will not continue the discussion of all elements of fiber drawing and film casting processes and refer the interested reader to the rich literature on these topics (e.g., Fourné 1999; Denn 2008; Agassant et al. 2017). Instead, both processes will now be simplified as much as possible so that we can focus on the main topic of this book, the draw resonance instability. As already noted above, the very general concepts of fiber drawing and film casting are similar: molten material is extruded through a die and, after flowing in an air gap, taken-up with a larger speed with the aim to make thin films or fibers. Figure 1.4 visualizes the reduced models and their similarities. Note that we assume at this point that in the entire air gap the material is in a molten state. For a fiber drawing process as depicted in Fig. 1.1, this means that we are reducing the analyzed domain to the channel. The most important parameter utilized to describe such forced elongation processes is the *draw ratio*, which will be denoted by Dr throughout this book and which is defined as the ratio of the velocity at the outlet to the one at the inlet,

$$Dr = \frac{u_{\text{outlet}}}{u_{\text{inlet}}} . \tag{1.1}$$

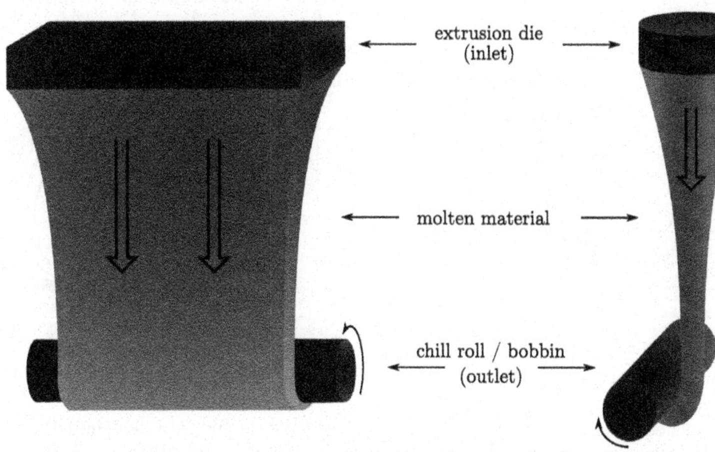

Fig. 1.4 Sketches of film casting (*left*) and fiber drawing (*right*). The flow direction is indicated by big arrows. A possible swelling effect close to the die is not shown

In general, large draw ratios are tried to be achieved in order to produce preferably thin films or fibers. There exists, however, a critical value of the draw ratio, denoted Dr_c, beyond which the formerly stable flow becomes unstable and both the velocity field and the fiber diameter or, respectively, the film thickness and width begin to oscillate: the *draw resonance* sets in. A first impression of this is given in Fig. 1.5 showing the transient evolution of the film thickness and the flow velocity at the middle of the casting length after an initial small perturbation is induced for stable and unstable processing conditions. In the stable regime, the initial perturbation leads to oscillations in both film thickness and axial velocity, but the amplitude of these oscillations decreases exponentially with time and the steady state is recovered (Fig. 1.5a, top), which is also represented by a shrinking spiral trajectory in the phase space until converging to a *fixed point* (Fig. 1.5a, bottom). For unstable conditions, the initially small perturbation exponentially grows in time and the oscillation amplitude finally saturates due to nonlinear effects to a finite value, leading to steady oscillations of thickness and velocity, represented by a closed loop in the phase space, also

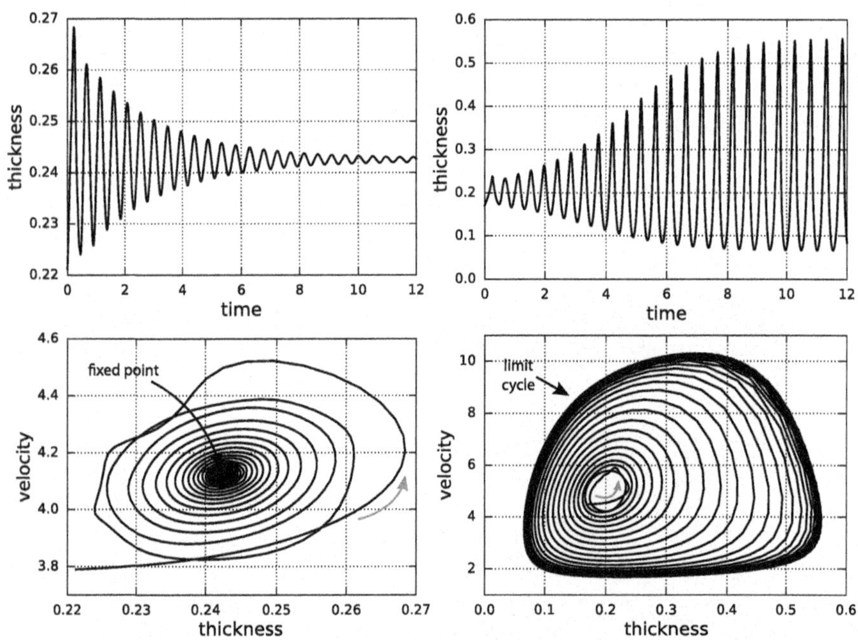

(a) Stable processing conditions: $Dr < Dr_c$ (b) Unstable processing conditions: $Dr > Dr_c$

Fig. 1.5 Transient response of the film thickness and the axial velocity on a small initial perturbation at the middle of the gap between the inlet and the outlet. **a** In the case of stable processing conditions, i.e., the draw ratio is below the critical value, the steady state (*fixed point*) is oscillatory reached. **b** In the case of unstable processing conditions, i.e., the draw ratio is above the critical value, the perturbation grows until permanent oscillations of constant amplitude (*limit cycle*) are established. The gray arrows in the phase space plots (bottom plots) indicate the time direction. Dimensionless variables are used

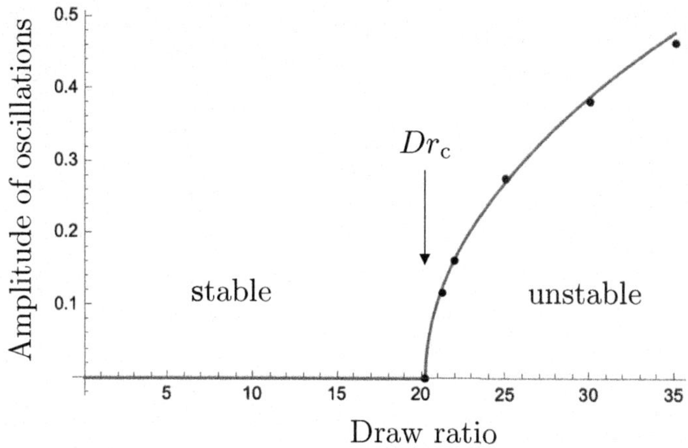

Fig. 1.6 Amplitude of permanent oscillations in draw resonance versus draw ratio Dr as obtained from numerical simulations (*black dots*). No oscillations are observed for $Dr < Dr_c$ for which the system is stable. Above the instability threshold for $Dr > Dr_c$, the amplitude is best fitted with $A = 0.1258\sqrt{Dr - Dr_c}$ (*solid line*)

referred to as a *limit cycle*. The width of the limit cycle corresponds to the amplitude of permanent oscillations that can be plotted as a function of the draw ratio, as shown in Fig. 1.6, and depicting a so-called *Hopf bifurcation*.

Figure 1.7 depicts the complete film thickness profile at various time steps. It can be more specifically seen that the instability manifests itself in waves of maximum and minimum thickness, which are alternatively traveling from the inlet to the outlet. As we will see later, those waves are mutually causing each other, which already represents the general mechanism behind the draw resonance effect. As the draw resonance leads to inhomogeneous thickness or even breakdown of the whole process, it is important to understand the instability mechanism and especially the dependence of the critical draw ratio on the variable process parameters. Keeping in mind the entire processes as presented in Sect. 1.1, this goal has been achieved step by step in literature by tuning separately the different effects and analyzing their influence on the draw resonance.

1.3 Literature Review

Starting from the first observations of draw resonance more than 60 years ago by Christensen (1962) and Miller (1963), the effect has been in the focus of a multitude of theoretical and experimental works.

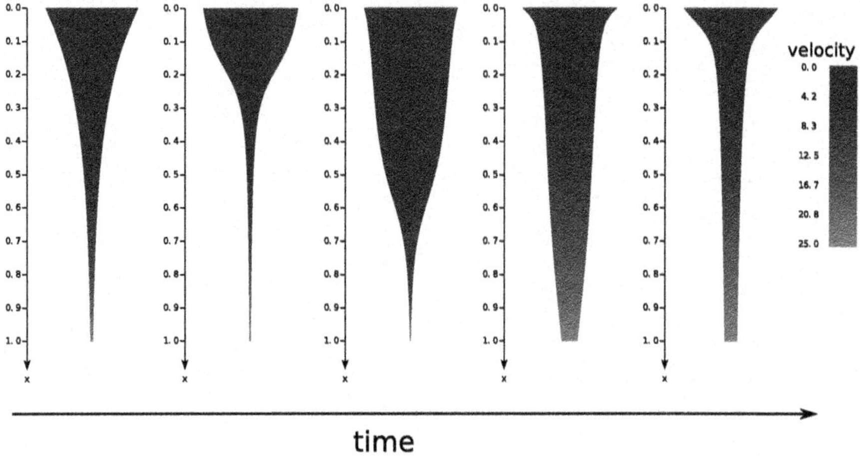

Fig. 1.7 Snapshots of the film thickness profile at different time steps over one oscillation period in the case of unstable processing conditions, i.e. $Dr > Dr_c$. The flow velocity values are coded in grey levels. Dimensionless variables are used

1.3.1 Theory

Historically, the analysis of the fiber drawing process was investigated earlier than the film casting process. The first works on the stability behavior were presented by Pearson and Matovich (1969), Gelder (1971) and Kase (1974).[4] All three works examine the linear stability of a purely viscous Newtonian fiber drawing process, i.e., neglecting gravity, inertia, surface tension, viscoelastic and non-isothermal effects, and are taken as the basis of many subsequent investigations. The value of the critical draw ratio was determined as 20.21, while modern calculations with higher accuracy yield a value of 20.218. Shah and Pearson (1972a) studied the influence of nonisothermal effects by varying the viscosity according to Arrhenius' law and stated a stabilizing effect of *spinline*[5] cooling. Afterwards, Shah and Pearson (1972b) extended the analysis to include the effects of gravity, inertia and surface tension using the Reynolds, Froude and Weber numbers as dimensionless parameters. While the critical draw ratio increased with increasing influence of inertia and gravity, surface tension led to a destabilization. These results were later confirmed qualitatively, but corrected quantitatively, by Chang et al. (1981). The simultaneous influence of inertia, gravity and surface tension effects is analyzed by Bechert and Scheid (2017). Chang et al. (1982) and Suman and Kumar (2009) also investigated the non-isothermal stability behavior of Newtonian fiber drawing.

[4] Within this work, the author refers to a previous work from 1966, which appeared in the Japanese journal *Seni Kikai Gakkaishi* and within which similar results are presented.

[5] The term "spinline" refers to the axial direction of the flow from the inlet to the outlet.

The first draw resonance study for the film casting process was performed by Yeow (1974). Assuming a film of infinite width[6] and small thickness led to a one-dimensional, purely viscous Newtonian model, which is mathematically equivalent to the simplest fiber model, as will be shown in Chap. 2. The assumption of small thickness, leading to independence of the flow variables across the thickness, is commonly referred to as the *membrane approximation*. Scheid et al. (2009a) extended Yeow's model to account for non-isothermal effects. Due to the typical geometry of large film widths compared to the thickness, surface tension effects are usually negligible. Extensions to cover the effects of inertia and gravity on the stability behavior in the one-dimensional case of infinite width are available from Cao et al. (2005) and Bechert et al. (2015). The work of Cao et al. (2005) is very similar to the one of German et al. (2006) for fibers due to the large similarities of both processes in the one-dimensional limit.

Besides linear stability analysis, there exist few works applying a weakly non-linear stability analysis (Fisher and Denn 1975; Schultz et al. 1984; Gupta and Chokshi 2015) and some more which explore the full transient solutions (Lee and Hyun 2001; Kim et al. 2005; Kallel et al. 2015). An alternative approach which deserves special attention are the kinematic wave studies introduced by Hyun (1978), later refined by Kim et al. (1996) and followed by several more (e.g., Jung et al. 1999, 2000; Lee et al. 2001b, 2005). The works of Renardy and Hagen (e.g., Renardy 2006; Hagen 2000; Hagen and Renardy 2000) provide analytical approaches to the stability behavior, including a solution for the complex perturbations and the asymptotic behavior of the eigenvalue spectrum. Detailed and specific literature overviews are provided in each chapter separately. It is striking that, despite the fact that the first studies have already been performed several decades ago, there still exist open questions related to draw resonance nowadays. A major reason for this is the enormous complexity of the problem as a whole, as it involves both a challenging flow problem with numerous influencing parameters and a non-linear viscoelastic material behavior, making a general steady state solution already a non-trivial task.

1.3.2 Experiments

The number of experimental investigations on draw resonance is rather small compared to the amount of theoretical work. It is mostly more difficult to analyze particular isolated effects in experiments as it is in theoretical investigations. Nevertheless, experimental data is mandatory to connect the theoretical results to an application in the real world. Experimental works related to film casting and fiber drawing are separately reviewed in this section.

In the case of film casting, the first experimental work on draw resonance was published by Barq et al. (1990), who produced films of polyethylene terephthalate

[6] Infinite width here means that the initial film width defined by the die geometry is much larger than the casting length between die and take-up, so that edge effects can be neglected.

(PET) with almost constant viscosity in the relevant deformation rate regime and subsequently measured the film thickness and width at various positions by sectioning the film with a grid. The results of three different resins were qualitatively compared by discussion of the thickness and width oscillation curves and miscellaneous perturbations were induced at the outlet. The study was repeated four years later using a PET with a larger relaxation time in order to account for viscoelastic effects and a small increase of the critical draw ratio was reported (Barq et al. 1994). Bouamra et al. (2006) presented a small study on film casting of four different polyethylenes (PE) with various relaxation times. The oscillations of the pulling force were measured and stability diagrams visualize the dependence of the critical draw ratio on the throughput. A threshold to unconditional stability for large throughputs was observed. In a mostly theoretical study, Shin et al. (2007) showed experimentally determined thickness oscillations for high-density and low-density PE films. Zavinska et al. (2008) performed film casting experiments using a commercial linear PE and determined the critical draw ratio for various values of the Deborah number[7] by varying the distance between inlet and outlet. Increasing the Deborah number led to a strong decrease of the critical draw ratio.

Probably the most comprehensive study was performed by Burghelea et al. (2012). They conducted simultaneous in-situ measurements of the width and the flow velocity field of PP films. The velocity field was measured by means of a digital particle image velocimetry. In the unstable regime, a limit cycle could be observed. The transition from the stable to the unstable regime of the velocity and width fluctuations was fitted with the stationary Landau–Ginzburg equation. Additionally, a third regime of chaotic behavior was reported for large draw ratios. The transition from stable to unstable behavior is listed for three different Weissenberg numbers,[8] revealing a destabilization with increasing Weissenberg number. Grieß (2014) measured the width variations in film casting of two linear and one long chain branched (LCB) polypropylene (PP). While draw resonance never occurred for the LCB PP, increasing the Deborah number always had a destabilizing effect for the linear materials.

In the case of fiber drawing, the first comprehensive experimental investigation on draw resonance had been performed by Bergonzoni and DiCresce (1966a, b). They extruded slender ribbons of PP and measured thickness and width oscillations. The transition to draw resonance was observed at draw ratios between 10 and 20, mostly close to 20. Additionally, the influence of temperature, die geometry and polymer properties on the oscillation amplitude was discussed. Donnelly and Weinberger (1975) were the first to observe draw resonance for a Newtonian fluid by using silicon oil. The theoretical value of $Dr_c \approx 20$ could be reproduced for various flow rates and the residence times were compared to the oscillation periods. The influence of strain hardening and thinning materials on draw resonance was studied by Weinberger et al. (1976) by using five different polymers, but no stabilizing effect could be observed

[7] The Deborah number compares the relaxation time of a material to the characteristic time of the process in which it is involved.

[8] The Weissenberg number compares the elastic forces to the viscous forces, which is equivalent to the Deborah number in this specific context.

for a branched low-density PE, although this material exhibits strain hardening. While the processing conditions were kept fix, a variation of the material made the investigation of various relaxation times possible. No clear trend of the critical draw ratio could be observed, but the data suggests rather a destabilizing influence of elasticity. Matsumoto and Bogue (1978) concentrated on the influence of the crystallization and solidification transitions induced by spinline cooling on draw resonance for polystyrene and PP. Chang and Denn (1979) performed experiments with both Newtonian and viscoelastic corn sirup solutions and measured the take-up force. The Newtonian critical draw ratio was determined to be around 25 and the viscoelastic value differed slightly, without revealing a clear trend behavior, though.

Kase (1974) and Blyler and Gieniewski (1980) reported that additional cooling can lead to the complete suppression of draw resonance. Ghijsels and Ente (1980) analyzed seven well characterized PPs of different properties and discussed the influence of molecular and processing parameters on the instability behavior. By proper material selection they could investigate the effects of various molecular weights and various molecular weight distribution widths separately. Both increasing the molecular weight and increasing the distribution width led to a decrease of the critical draw ratio. The drawing distance was found to influence the amplitude of oscillation, but not the critical draw ratio. A larger critical draw ratio was detected with decreasing mass flow rate. Nam and Bogue (1984) performed isothermal experiments with two PPs and varied the die variables, extrusion speed and spinning length. They observed a stability minimum with respect to the inlet velocity. In another work, Demay and Agassant (1985) studied the influence of mass flow rate, take-up speed and spinning length on draw resonance for four different polyesters of both linear and branched type. The critical draw ratio increased with increasing Stanton number[9] and with increasing spinning length. A threshold length above which no draw resonance could be observed is reported. Unfortunately, the presented data is not enough to analyze a potential correlation between the Deborah number and the critical draw ratio for a particular material. The results of White and Yamane (1985) for several linear and branched PETs coincide with the other findings described above. Additionally, they mention a possible stabilizing effect due to strain hardening.

1.3.3 Reviews

There exist also several review articles on draw resonance. The first comprehensive one is given by Petrie and Denn (1976) and covers both experimental and theoretical findings. Denn (1980a) and Larson (1992) shortly review draw resonance within the framework of more general reviews. Newer articles are given by Hyun (1999), who provides a short chronological overview and Jung and Hyun (2006), who summarize experimental and theoretical investigations on fiber spinning, film casting and film

[9] The Stanton number compares the heat transferred to/from the fluid to the heat transported by the fluid.

blowing. Another comprehensive introduction and summary can be found in Chap. 10 by Co (2004) and Chap. 11 by Jung (2004) of the textbook edited by Hatzikiriakos and Migler (2004). The most recent, however rather short, review on draw resonance in fiber spinning, film casting and film blowing is available from Demay and Agassant (2014).

1.4 Observing the Critical Draw Ratio and Frequency

Even though there are several experimental data available, only few of them can be used to validate theoretical predictions. A drawback of many investigations is actually the imprecise method to determine the critical draw ratio, which is sometimes evaluated based on fibers or films after the process, where additional effects could have occurred in the meantime, or by simply observing the material by eye and estimating the critical value to lie between a stable and an unstable setting. The latter method, however, is prone to errors, as the transition from the stable to the unstable regime is imperfect due to background noise. As a consequence, the critical draw ratio could lie above a value for which oscillations may become visible already.

We present below measurement results that use the in-situ method of Burghelea et al. (2012) to determine the critical draw ratio in fiber drawing experiments performed with a linear polypropylene (PP1). The fiber geometry has the advantage that the neck-in effect of film casting can be ignored, and the parameters have been chosen such that inertia, gravity and surface tension are negligible as well, which makes it possible to concentrate primarily on viscoelastic effects.

Figure 1.8 shows the image processing sequence for observing draw resonance, with r the fiber radius. The time dependence of the determined fiber diameter at the fifth position, $2r^{(4)}$, is plotted for three different draw ratios, below criticality (c1), close to criticality (c2), and above (c3). Below criticality, merely non-periodic noise is visible, while clear oscillations in diameter become observable for larger draw ratios. The oscillation amplitude increases with the draw ratio.

For determining the critical draw ratio, the method proposed by Burghelea et al. (2012) is adopted here. The reduced mean square fluctuation of the diameter Ξ, defined by

$$\Xi = \frac{< (2r- < 2r >)^2 >}{< 2r >}, \tag{1.2}$$

with $< \cdot >$ denoting the mean value with respect to time, is used as a measure for the instability. Figure 1.9 shows this quantity at the fifth measurement position of the fiber for various draw ratios. While Ξ is rather constant for low values of Dr, it starts to increase up to more than one order of magnitude for Dr above a critical value. This critical draw ratio Dr_c, however, is not clearly detectable at first glance. The reason for this is the presence of background noise, which blurs the ideally sharp transition from stable to unstable behavior. If no noise would be present, such

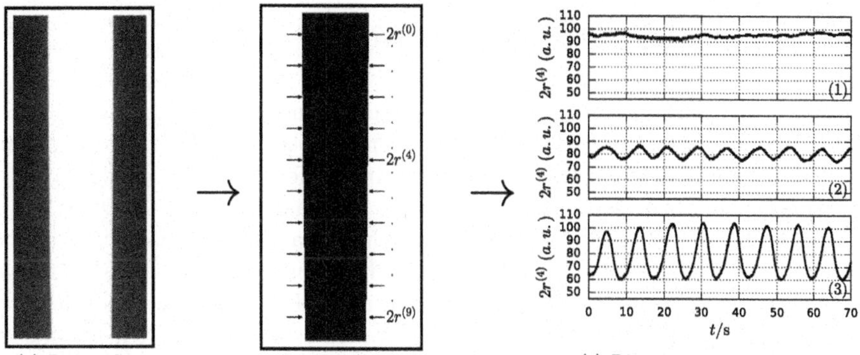

(a) Raw video. (b) Binary video. (c) Diameter measurement.

Fig. 1.8 Visualization of the video post-processing steps, exemplarily for a linear polypropylene (PP1) fiber with a spinning length of 20 cm and an inlet speed of 31 mm/s. The raw video (**a**) is converted into inverted binary frames (**b**) and the diameter of the fiber in arbitrary units (a.u.) is measured at 10 equidistant heights. Draw resonance oscillations become clearly visible at the fifth position with increasing draw ratio with (**c**1) $Dr = 1.3$, (**c**2) $Dr = 5.2$ and (**c**3) $Dr = 7.7$

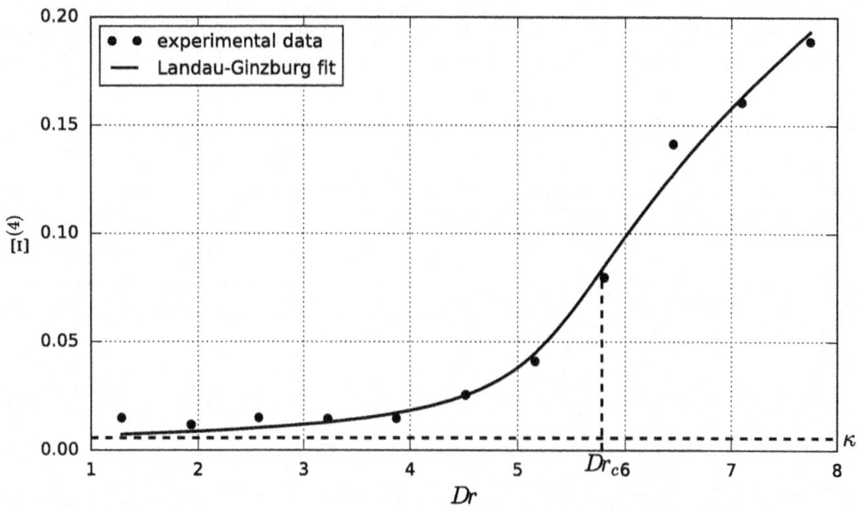

Fig. 1.9 Evolution of the reduced mean square diameter fluctuation Ξ at the fifth position of the fiber with increasing draw ratios, for the same experimental parameters than in Fig. 1.8. The transition from the stable to the unstable regime can be fitted with the stationary Landau–Ginzburg equation (1.3), which makes a defined determination of the critical draw ratio possible

bifurcations could be described in the vicinity of the critical parameter by a square root function in the unstable regime (Strogatz 2000) as shown in Fig. 1.6. Here, a description of an *imperfect Hopf bifurcation* is needed, which can be done with the stationary Landau–Ginzburg equation (Landau and Lifshitz 1980), written as

$$\left(\frac{Dr}{Dr_c} - 1\right) \Xi - \beta\, \Xi^3 + \kappa = 0\,. \tag{1.3}$$

Besides the critical draw ratio Dr_c, two parameters β and κ tune the transition behavior. While β quantifies the inverse growth rate in the unstable regime, κ is an external field, i.e., a measure for the background noise. In fact, for $\kappa \to 0$, (1.3) reduces to the previously described square root form for $Dr > Dr_c$ and to $\Xi = 0$ for $Dr < Dr_c$.

Figure 1.9 shows the fitting of the experimental data using (1.3) together with the marks of the resulting critical draw ratio Dr_c and the noise parameter κ (dashed lines). At this point it becomes clearly visible that the critical draw ratio determined this way can be significantly different from a value obtained by a more simple method, e.g., choosing the smallest value of Dr at which oscillations become visible. The fitting procedure was applied at all positions in height separately and the mean value of the critical draw ratio was evaluated afterwards. The inverse growth rate β was found to decrease with increasing x-value, i.e., the oscillation amplitude increases towards the outlet. A linear stability analysis does not yield any information about the dynamics in the vicinity of the critical point, though, and parameter β will not be examined further in the scope of this book. The advantages of the method used here are that it is an objective procedure and that the resulting critical draw ratio directly corresponds to the theoretical value calculated in the linear stability analysis, as additional noise, which is not covered by the ideal model, is implicitly respected.

The oscillation frequency at criticality f_c can be determined by applying a fast Fourier transformation (FFT) on the transient fiber diameter data. Figure 1.10 shows both the diameter with respect to time t and the corresponding frequency spectrum with respect to frequency f for two draw ratios, one belonging to stable conditions and one close to the critical value. A peak in the spectrum can be detected in the latter case, located at the frequency f_c of the purely sinusoidal oscillation present at

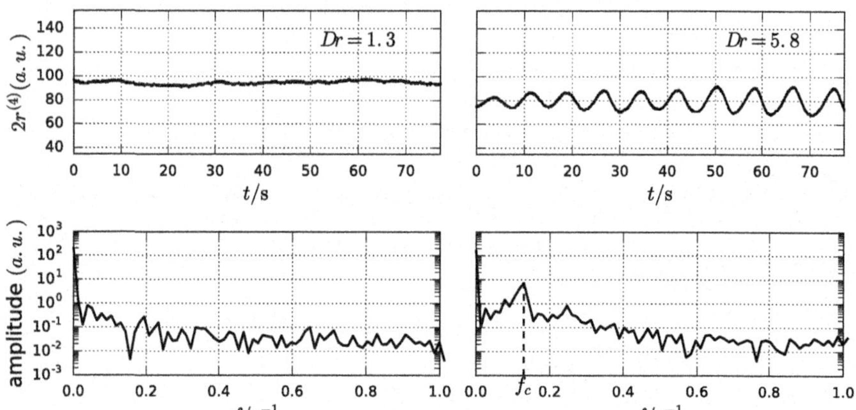

Fig. 1.10 Transient fiber diameter (fifth position, see Fig. 1.8) and the corresponding frequency spectrum obtained via FFT for $Dr = 1.3$ (*stable*) and $Dr = 5.8$ (*critical*)

criticality. Due to experimental limitations, it is not always possible to operate exactly at critical conditions and the data set of the draw ratio closest to the critical value was chosen for the frequency analysis. In the demonstration case here, the available data is very close to the critical value, as Fig. 1.9 illustrates. In any case, the variation of the frequency in the vicinity of the critical draw ratio are usually very small. The fact that the oscillation frequency seems to be rather independent of the draw ratio as well as on the different position in height also indicates that the observed frequency may be the result of an oscillation forced by periodic noise, e.g., by a non-constant throughput rate of the extruder. This again illustrates the complexity of observing draw resonance in experiments and plead in favor of thoroughful theoretical analysis, which is the topic of this book.

1.5 Modeling Draw Resonance and Outline

Every approach of modeling in materials processing is based on two types of equations, namely the *governing equations* and the *constitutive equations*. The former group consists of fundamental laws of physics, which are valid independently of the choice of material. In the context of this book, those equations are the conservation of mass,

$$\partial_t \rho + \nabla \cdot (\rho \, \mathbf{v}) = 0 \, , \tag{1.4}$$

the conservation of momentum,

$$\rho \, (\partial_t \mathbf{v} + (\mathbf{v} \cdot \nabla)\mathbf{v}) = \nabla \cdot \sigma + \rho \, \mathbf{b} \, , \tag{1.5}$$

and the conservation of energy, neglecting the viscous heating,

$$\rho c_p \, (\partial_t T + \mathbf{v} \cdot \nabla T) = \nabla \cdot (k \nabla T) \, , \tag{1.6}$$

with time t, density ρ, velocity field $\mathbf{v}$, stress tensor σ, body force vector $\mathbf{b}$, specific heat capacity c_p, temperature field T, and heat conductivity k. The nabla operator ∇ is defined as $\nabla = (\partial_x, \partial_y, \partial_z)$ and $\partial_i = \partial / \partial i$ is used as abbreviation for all partial derivatives throughout this book.

The selection of the set of governing equations depends on the chosen assumptions for the process model. As the system of equations in fluid dynamics is usually of high complexity, it is common to reduce the equations to a minimum by eliminating terms of negligible contribution. On the other hand, accounting for particular effects, e.g., inertia or gravity, introduces dependencies on practically tunable and measurable parameters, e.g., extrusion and take-up velocities, casting length or melt and ambient temperatures, all entering in boundary conditions. We will refer to those parameters, which describe the process without accounting for material properties, as *process parameters*.

In order to link a particular material behavior to the governing equations, a constitutive equation is needed for the conservation of momentum,[10] which describes the relationship between deformation and external stresses. The simplest example of such an equation is the well-known *Newtonian fluid model*, which can be written as

$$\tau = 2\,\eta\,\mathbf{D}\,, \tag{1.7}$$

with the rate-of-deformation tensor

$$\mathbf{D} = \frac{1}{2}\left((\nabla\mathbf{v}) + (\nabla\mathbf{v})^T\right)\,. \tag{1.8}$$

Here, the extra-stress tensor $\tau = p\,\mathbf{I} + \sigma$ is used, where p denotes the isotropic pressure and $\mathbf{I}$ the identity matrix. A Newtonian fluid is entirely characterized by its shear viscosity η, which is a measure for the resistance to external stresses.

In contrast to Newtonian fluids, to which molten glass is often associated, polymer melts belong to the group of viscoelastic fluids, which makes (1.7) often an inappropriate choice. There exist a multitude of viscoelastic constitutive equations, among which the most simple one is the so-called *upper-convected Maxwell (UCM)* equation, given as

$$\lambda\,\overset{\nabla}{\tau} + \tau = 2\,\eta\,\mathbf{D}\,, \tag{1.9}$$

with the upper-convected time derivative

$$\overset{\nabla}{\tau} = \partial_t\tau + \mathbf{v}\cdot\nabla\tau - \tau\cdot\nabla\mathbf{v} - (\nabla\mathbf{v})^T\cdot\tau\,. \tag{1.10}$$

Equation (1.9) introduces the relaxation time λ, which defines the intrinsic time scale of the material within which elastic processes occur. We will refer to parameters like the viscosity, relaxation time, density, heat capacity and heat conductivity as *material parameters*.

The central question in studying draw resonance can now be formulated as follows: *Which process and material parameters influence the draw resonance behavior, how, and under which circumstances?*

To this aim, we present in **Chap.** 2 the one-dimensional Newtonian model for both fiber drawing and film casting in isothermal conditions, emphasizing the similarities and the differences between each model. **Chap.** 3 gives the formalism on the linear stability analysis that will be applied in subsequent chapters, as well as more insights on the physical mechanism of draw resonance. The next three chapters concentrate on effects which are independent of a particular constitutive equation. For this purpose, Newtonian models are employed exclusively. In particular, the influences on draw

[10] Note that the Fourier's law that has been used to write the conservation of energy is also a constitutive equation, but that will be used as is throughout the entire book and not subject to discussion then.

resonance of gravity and inertia are treated in **Chap.** 4. The effect of surface tension only relevant in fiber drawing is analyzed in **Chap.** 5, while the neck-in effect only relevant in film casting is investigated in **Chap.** 6.

Afterwards, viscoelastic effects are investigated in the next two chapters. As there are many different viscoelastic constitutive equations available, **Chap.** 7 gives an overview of several prominent examples. In contrast to the previous chapter, additional effects like gravity and inertia are neglected in order to focus on the pure influence of viscoelasticity. The different models are derived in **Chap.** 7 and then analyzed in **Chap.** 8 with respect to their stability behavior, which is followed by a compact evaluation of the suitability of each constitutive equation. The purpose here is to provide a systematic overview of various viscoelastic fiber drawing and film casting models to support the search of the most appropriate one for a particular application. In addition, we will look at the viscoelastic stability mechanism and build a bridge to well-known viscoelastic effects on the viscous flow behavior, e.g., strain hardening or thinning. This includes the study of generalized Newtonian fluid models, which are described by the Newtonian equation (1.7), but with the viscosity depending on the fluid deformation.

Chapters 4 to 8 are rather fundamental, isolated studies on single process and material effects. In reality, these effects usually superimpose, and models accounting for several effects simultaneously are needed for a proper description. **Chapter** 9 deals with the analysis of such models. In order to stick close to practical application, an alternative scaling is proposed in the beginning of the chapter, which leads to dimensionless variables with a one-to-one relation to the typically variable control parameters in practice. Newtonian and viscoelastic film and fiber models are treated and the stability behavior is visualized by stability maps, which yield a compact overview with respect to the practical control parameters and enable additionally a division of the parameter space into various regions, within which one or several effects can be neglected, as for instance parameter regions of unconditional stability in the case of dominating inertia or unconditional instability in the case of dominating surface tension in fiber drawing. When possible, easy solvable approximative stability criteria will be derived.

As already indicated in the beginning, thermal effects, especially cooling, often plays an important role in film casting and fiber drawing processes. **Chapters** 10 and 11 concentrate on the influence of the ambient temperature on draw resonance in film casting and fiber drawing, respectively. After including the energy equation to our models, we will investigate the change in stability threshold due to constant and variable, well-defined ambient temperature profiles. It is shown how heat transfer can strongly stabilize the processes, allowing in practice very high draw ratios yet in stable conditions. Conclusions and perspectives are given in **Chap.** 12.

Chapter 2
One-Dimensional Newtonian Models

Abstract This chapter presents the framework in which models of reduced dimensionality can be obtained for both the two-dimensional fiber drawing and the axisymmetric film casting configurations. The main assumptions are first introduced, followed by all steps necessary to derive the one-dimensional fiber and film models, starting from the two-dimensional mass and momentum conservation equations for isothermal and Newtonian fluids. The key feature is the use of an asymptotic expansion of the field variables based on the smallness of the aspect ratio of fibers and films. The derivations are made in a consistent way allowing to identify the main difference between both models, which only consists in a surface tension term, only present in the fiber model. The models are actually identical otherwise, safe the Trouton ratio, a constant prefactor entering into the extensional viscosity.

Due to the axisymmetric setup of the fiber drawing process, models can generally be reduced to two dimensions, i.e., an axial and a radial direction. We will focus here exclusively on slender fibers, which means that the radius of the fiber is small compared to the drawing length. As will be shown below, the axial velocity does not vary with the radius at leading order in this slender fiber approximation and a one-dimensional model can be used. The most simple film casting model is one-dimensional as well. In this case, it is assumed that the width of the film is very large compared to the casting length and thus infinite, eliminating one of initially three relevant spatial dimensions. Applying additionally the so-called *lubrication approximation*,[1] i.e., assuming that the film thickness is small compared to the casting length, results in a one-dimensional model, as the axial velocity does not vary across the thickness at leading order. The one-dimensional fiber model used in this book is based on the model of Matovich and Pearson (1969) and further generalizations (e.g., Shah and Pearson 1972b), while the one-dimensional film casting model follows the approach of Yeow (1974) and succeeding works (e.g., Cao et al. 2005). More details on the derivation of fiber models are available in Schultz and Davis (1982), Dewynne et al. (1989).

[1] Sometimes referred to as the *membrane approximation*, but in the context of films only, while *lubrication approximation* applies for both films and fibers.

© Springer Nature Switzerland AG 2025

M. Bechert and B. Scheid, *Draw Resonance*, Fluid Mechanics and Its Applications 141,
https://doi.org/10.1007/978-3-031-85568-9_2

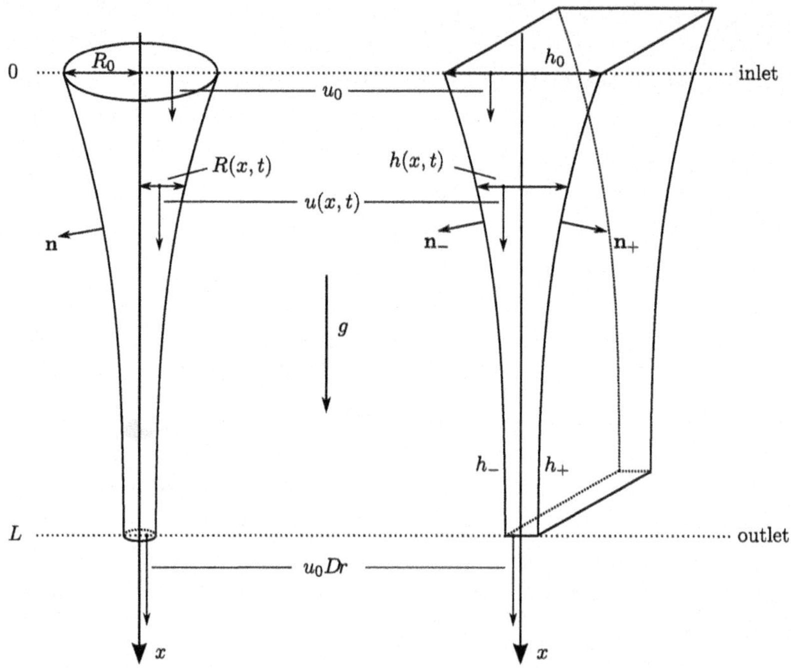

Fig. 2.1 Sketches of the one-dimensional models for fiber drawing (*left*) and film casting (*right*)

Figure 2.1 shows the sketches of one-dimensional models for both fiber drawing and film casting. The following variable system is used to describe the processes. The flow velocity in x direction is denoted by u, R is the fiber radius and h the film thickness. The drawing length or, respectively, the casting length is denoted by L. The initial velocity, radius and thickness are indicated by the subscript '0' and are practically defined by the extrusion velocity and the die geometry. For the sake of simplicity, u_0 denotes hereby the cross-section-averaged initial velocity. Later, we will see that the axial velocity does not vary over the cross section in the one-dimensional models. For this reason, the averaged value at the inlet is sufficient. The ratio of outlet to inlet velocity is named the *draw ratio* Dr, such that the fixed (cross-section averaged) outlet velocity at the take-up is equal to $u_0 Dr$. As can be seen in Fig. 2.1, the similarities between the one-dimensional models of fiber drawing and film casting are numerous. Nevertheless, there are differences due to the geometrical forms. While surface tension effects must not be neglected generally in fiber drawing due to the circular shape of the fiber, geometrical effects like neck-in will come into play in film casting of finite-width films. A possible way to extend the model to cover such finite-width effects is presented in Chap. 6.

Let us summarize all assumptions of the one-dimensional Newtonian models:

i The slender fiber or, respectively, membrane approximation is valid, i.e., $R/L \ll 1$ or, respectively, $h/L \ll 1$,
ii The material density is constant,
iii The effect of air drag is neglected,
iv The process is running under isothermal conditions,
v The material behavior is governed by the Newtonian constitutive equation (1.7),

and, in the case of film casting,

vi The film is of infinite width.

The first assumption (i) is crucial and at the heart of all one-dimensional models, as it enables us to reduce the dimensionality of the equations, which will be shown in the next section. The assumption of constant material density (ii) is a reasonable choice for many processes as the density typically depends only weakly on the temperature, provided that no phase transition, e.g., from liquid to solid state, occurs. Similarly, the effect of air drag can safely be neglected in many cases (iii), as long as there is no air flow added from the outside, like the cooling air sketched in Fig. 1.1. The remaining three assumptions (iv,v,vi) will be dropped in the scope of later chapters: finite-width effects are taken into account in Chap. 6, non-Newtonian viscoelastic models are introduced in Chap. 7 and analyzed in Chap. 8, and non-isothermal processes are treated in Chap. 10 for film casting and in Chap. 11 for fiber drawing.

2.1 Fiber Model

We will use cylindrical coordinates for the description of fiber drawing. The flow field is assumed to be axisymmetric and thus given by $\mathbf{v} = u\,\mathbf{e}_x + v\,\mathbf{e}_r$, where $\mathbf{e}_x$ and $\mathbf{e}_r$ are the unit vectors in axial and radial directions, respectively, and u and v are the corresponding velocity field components. The fiber surface is at position $r = R(x,t)$, such that the outward vector normal to it is given by $\mathbf{n} = (-\partial_x R\,\mathbf{e}_x + \mathbf{e}_r)/\sqrt{1+(\partial_x R)^2}$. Due to the axisymmetric setup, all derivatives of the variables with respect to the angular variable φ can be neglected.[2] The two principal radii of curvature at the interface are given by basic differential geometry calculus as $R_1 = \left(1+(\partial_x R)^2\right)^{3/2}/\partial_{xx}R$ and $R_2 = -R\sqrt{1+(\partial_x R)^2}$ (Dantzig and Tucker 2001).

The flow is described by the continuity equation (1.4), which simplifies due to the constant density (assumption ii) to

$$\nabla \cdot \mathbf{v} = 0\,, \tag{2.1}$$

and by the momentum conservation,

[2] It has to be noticed, though, that $\partial_\varphi \mathbf{e}_\varphi = -\mathbf{e}_r$.

$$\rho(\partial_t \mathbf{v} + \mathbf{v} \cdot \nabla \mathbf{v}) = \nabla \cdot \sigma + \rho \mathbf{g} \,, \tag{2.2}$$

with $\nabla = \mathbf{e}_r\,\partial_r + \mathbf{e}_\varphi\,\partial_\varphi/r + \mathbf{e}_x\,\partial_x$, $\mathbf{g} = g\,\mathbf{e}_x$, gravitational acceleration g, material density ρ and stress tensor $\sigma = -p\,\mathbf{I} + \tau$, where p denotes the pressure, and the extra-stress tensor τ is given by the Newtonian constitutive equation (1.7). At the fiber surface, one needs to fulfill the kinematic boundary condition,

$$\partial_t R + u|_R \partial_x R - v|_R = 0 \,, \tag{2.3}$$

where $\cdot|_R$ indicates that the variable is taken at the boundary in $r = R$, and the stress boundary condition, accounting for the Laplace pressure at the interface,

$$\mathbf{n} \cdot \sigma|_R = \gamma \left(R_1^{-1} + R_2^{-1}\right) \mathbf{n} \,, \tag{2.4}$$

with γ the surface tension. Note that the viscous stress in (2.4) from the surrounding air has been neglected (assumption iii) and its pressure has been set to zero without loss of generality.

The variables are made dimensionless using the following transformation rules:

$$\begin{aligned} x \to L\,x \,,\ \ r \to \alpha\,L\,r \,,\ \ R \to \alpha\,L\,R \,,\ \ u \to u_0\,u \,,\ \ v \to \alpha\,u_0\,v \,, \\ t \to \frac{L}{u_0} t \,,\ \ \sigma \to \eta\,\frac{u_0}{L}\,\sigma \,,\ \ p \to \eta\,\frac{u_0}{L}\,p \,, \end{aligned} \tag{2.5}$$

where the fiber parameter $\alpha = R_0/L$ is used. The dimensionless numbers resulting from this scaling are the Reynolds number Re, the Froude number Fr and the rescaled capillary number Ca^*, defined as

$$Re = \frac{\rho\,u_0\,L}{\eta} \,,\ \ Fr = \frac{u_0^2}{g\,L} \,,\ \ Ca^* = \alpha\,Ca = \frac{\alpha\,\eta\,u_0}{\gamma} \,, \tag{2.6}$$

with the commonly used definition of the capillary number Ca. We will refer to Ca^* merely as the "capillary number" within this book. The scaled continuity equation (2.1) then becomes

$$\frac{1}{r}\partial_r(rv) + \partial_x u = 0 \,, \tag{2.7}$$

and the axial and radial components of the momentum conservation (2.2) are

$$\alpha^2 Re(\partial_t u + u\partial_x u + v\partial_r u) = \alpha^2 \partial_x \sigma_{xx} + \frac{1}{r}\partial_r(r\sigma^*_{rx}) + \alpha^2 \frac{Re}{Fr} \,, \tag{2.8a}$$

$$\alpha^2 Re(\partial_t v + u\partial_x v + v\partial_r v) = \frac{1}{r}\partial_r(r\sigma_{rr}) + \partial_x \sigma^*_{rx} - \frac{\sigma_{\varphi\varphi}}{r} \,, \tag{2.8b}$$

where $\sigma^*_{rx} := \alpha\sigma_{rx}$ and $\sigma_{\varphi\varphi} = -p + 2v/r$. The kinematic boundary condition (2.3) is not changed by the scaling transformation. The axial and radial components of the stress boundary condition (2.4) can be written as

$$\sigma^*_{rx}|_R - \alpha^2\sigma_{xx}|_R\partial_x R = -\frac{\alpha^2}{Ca^*}\partial_x R\,\frac{-1+\alpha^2\left(R\partial_{xx}R - (\partial_x R)^2\right)}{R\left(1+\alpha^2(\partial_x R)^2\right)^{3/2}}\,, \tag{2.9a}$$

$$\sigma_{rr}|_R - \sigma^*_{rx}|_R\partial_x R = \frac{1}{Ca^*}\frac{-1+\alpha^2\left(R\partial_{xx}R - (\partial_x R)^2\right)}{R\left(1+\alpha^2(\partial_x R)^2\right)^{3/2}}\,. \tag{2.9b}$$

As the fiber parameter is assumed to be small (assumption i), all field variables will be expanded in the order parameter α^2, i.e.,

$$u = u^{(0)} + \alpha^2 u^{(1)} + \mathcal{O}(\alpha^4)\,, \tag{2.10}$$

$$v = v^{(0)} + \alpha^2 v^{(1)} + \mathcal{O}(\alpha^4)\,, \tag{2.11}$$

$$p = p^{(0)} + \alpha^2 p^{(1)} + \mathcal{O}(\alpha^4)\,, \tag{2.12}$$

$$\sigma = \sigma^{(0)} + \alpha^2 \sigma^{(1)} + \mathcal{O}(\alpha^4)\,. \tag{2.13}$$

These expansions are then plugged in (2.7), (2.8), and (2.9). Due to the scaling, the dimensionless variables are at the order of unity, noted $\mathcal{O}(1)$. As a direct consequence, a term which is, for instance, proportional to α^2 cannot influence a term which is proportional to α^0, i.e., 1, as α is small. This is the key feature of such an expansion, which enables us to split the expanded equations into multiple, independent system of equations at each order. At leading order, i.e., $\mathcal{O}(\alpha^0)$, we obtain the following system of equations:

$$\frac{1}{r}\partial_r\left(r\,v^{(0)}\right) + \partial_x u^{(0)} = 0\,, \tag{2.14a}$$

$$0 = \frac{1}{r}\partial_r\left(r\sigma^{*(0)}_{rx}\right)\,, \tag{2.14b}$$

$$0 = \frac{1}{r}\partial_r\left(r\sigma^{(0)}_{rr}\right) + \partial_x\sigma^{*(0)}_{rx} - \frac{\sigma^{(0)}_{\varphi\varphi}}{r}\,, \tag{2.14c}$$

$$\sigma^{*(0)}_{rx}|_R = 0\,, \tag{2.14d}$$

$$\sigma^{(0)}_{rr}|_R - \sigma^{*(0)}_{rx}|_R\partial_x R = -\frac{1}{Ca^*}\frac{1}{R}\,, \tag{2.14e}$$

$$\partial_t R + u^{(0)}|_R\partial_x R - v^{(0)}|_R = 0\,. \tag{2.14f}$$

Equations (2.14b) and (2.14d) directly yield

$$\sigma^{*(0)}_{rx} = \partial_r u^{(0)} = 0\,. \tag{2.15}$$

Consequently, the axial velocity does not vary along the radius at leading order, i.e., $u^{(0)} = u^{(0)}(x, t)$, which is referred to the *extensional velocity*. This makes it an easy task to average the continuity equation (2.14a) over the cross-section by integrating both sides with $\int_0^R r dr \int_0^{2\pi} d\varphi/(\pi R^2)$. Using in addition the kinematic condition (2.14f), the averaged continuity equation can be written as

$$\partial_t a + \partial_x \left(a u^{(0)}\right) = 0 \,, \tag{2.16}$$

with scaled fiber cross-section $a = \int_0^R r dr \int_0^{2\pi} \frac{d\varphi}{\pi} = R^2$. Moreover, we can exploit the fact that $u^{(0)}$ is independent of r to integrate (2.14a) to obtain

$$v^{(0)} = -\frac{r}{2}\partial_x u^{(0)} \,, \tag{2.17}$$

in which the axisymmetric condition $v|_{r=0} = 0$ was used.

Utilizing (2.17), the normal stress difference $\nu = \tau_{xx} - \tau_{rr}$ can be calculated at leading order as

$$\nu^{(0)} = 3\, \partial_x u^{(0)} \,, \tag{2.18}$$

where the factor 3 is the so-called *Trouton ratio* for purely *uniaxial deformation*. This means the external, deforming stresses act exclusively along one axis, which is then associated to deformation of the lateral boundaries. Moreover, using (2.17), it can be shown that $\sigma_{rr}^{(0)} = -p^{(0)} + 2\, \partial_r v^{(0)}$ is independent of r, and therefore $\sigma_{rr}^{(0)}|_R = \sigma_{rr}^{(0)}$. The pressure at leading order can therefore be calculated with the help of (2.14e) and (2.15) as

$$p^{(0)} = \frac{1}{Ca^* R} - \partial_x u^{(0)} \,, \tag{2.19}$$

and using (2.18) yields

$$\sigma_{xx}^{(0)} = \nu^{(0)} - \frac{1}{Ca^* \sqrt{a}} \,. \tag{2.20}$$

We have not used (2.14c) yet, which, based on the already gained information, gives

$$\sigma_{\varphi\varphi}^{(0)} = \sigma_{rr}^{(0)} \,. \tag{2.21}$$

However, this step is not needed to close the one-dimensional model and is merely given for the sake of completeness here. Instead, in order to proceed further, it is necessary to investigate the α^2-order of (2.8a) and (2.9a), which is given by:

$$Re\left(\partial_t u^{(0)} + u^{(0)}\partial_x u^{(0)}\right) = \partial_x \sigma_{xx}^{(0)} + \frac{1}{r}\partial_r\left(r\sigma_{rx}^{*(1)}\right) + \frac{Re}{Fr} \,, \tag{2.22a}$$

$$\sigma_{rx}^{*(1)}|_R - \sigma_{xx}^{(0)}|_R \partial_x R = \frac{1}{Ca^*} \frac{\partial_x R}{R} . \tag{2.22b}$$

Cross-section averaging (2.22a) leads to

$$Re \left(\partial_t u^{(0)} + u^{(0)} \partial_x u^{(0)} \right) = \partial_x \sigma_{xx}^{(0)} + \frac{2}{R} \sigma_{rx}^{*(1)}|_R + \frac{Re}{Fr} , \tag{2.23}$$

and we can substitute (2.22b) to eliminate the higher order term, which finally leads to

$$Re\, a \left(\partial_t u^{(0)} + u^{(0)} \partial_x u^{(0)} \right) = \partial_x \left(a \nu^{(0)} \right) + \frac{Re}{Fr} a + \frac{1}{2Ca^*} \frac{\partial_x a}{\sqrt{a}} . \tag{2.24}$$

The system of Eqs. (2.16), (2.18) and (2.24) is completed by the boundary conditions induced by the process, i.e.,

$$a(x = 0, t) = u(x = 0, t) = 1 , \tag{2.25a}$$

$$u(x = 1, t) = Dr , \tag{2.25b}$$

and determines the evolution of the cross-section and the axial velocity of the fiber drawing process. Note that the superscripts "(0)" will be omitted in the subsequent chapters. This model can easily be reduced to several limiting cases, as inertia, gravity and/or surface tension effects are neglected if Re, Re/Fr and/or $1/Ca^*$ are set to zero, respectively. If all three control parameters become zero, a purely viscous extensional flow is obtained.

2.2 Film Model

The derivation of the one-dimensional film casting model is very similar to the fiber model above, but a Cartesian coordinate system is used. As the film width is assumed to be infinite, only two spatial dimensions are left and $\nabla = \mathbf{e}_x\, \partial_x + \mathbf{e}_y\, \partial_y$. The flow field is then given by $\mathbf{v} = u\, \mathbf{e}_x + v\, \mathbf{e}_y$, where $\mathbf{e}_x$ and $\mathbf{e}_y$ are the unit vectors in casting and thickness direction, respectively, and u and v are the corresponding velocity field components. The normal vectors at the two surfaces $h_\pm = \pm 1/2\, h$ are $\mathbf{n}_\pm = (-\partial_x h_\pm \mathbf{e}_x + \mathbf{e}_y)/\sqrt{1 + (\partial_x h)^2/4}$.

The kinematic boundary condition reads

$$\partial_t h_\pm + u|_{h_\pm} \partial_x h_\pm - v|_{h_\pm} = 0 , \tag{2.26}$$

and the stress boundary condition simplifies to

$$\mathbf{n}_\pm \cdot \sigma|_{h_\pm} = 0 , \tag{2.27}$$

as surface tension effects can be neglected.[3] The scaling rules are given by

$$x \to L\,x\,,\quad y \to \beta\,L\,y\,,\quad h \to \beta\,L\,h\,,\quad u \to u_0\,u\,,\quad v \to \beta\,u_0\,v\,,$$
$$t \to \frac{L}{u_0}t\,,\quad \sigma \to \eta\frac{u_0}{L}\sigma\,,\quad p \to \eta\frac{u_0}{L}p\,, \tag{2.28}$$

with the film parameter $\beta = h_0/L$. The relevant control parameters are Re and Re/Fr as defined by (2.6). The scaled continuity equation can be written as

$$\partial_x u + \partial_y v = 0\,, \tag{2.29}$$

and the x- and y-components of the momentum conservation (2.2) are

$$\beta^2 Re(\partial_t u + u\partial_x u + v\partial_y u) = \beta^2\partial_x\sigma_{xx} + \partial_y\sigma^*_{xy} + \beta^2\frac{Re}{Fr}\,, \tag{2.30a}$$
$$\beta^2 Re(\partial_t v + u\partial_x v + v\partial_y v) = \partial_x\sigma^*_{xy} + \partial_y\sigma_{yy}\,, \tag{2.30b}$$

where $\sigma^*_{xy} := \beta\sigma_{xy}$. The x- and y-components of the stress boundary condition (2.27) are

$$-\beta^2\partial_x h_\pm\sigma_{xx}|_{h_\pm} + \sigma^*_{xy}|_{h_\pm} = 0\,, \tag{2.31a}$$
$$-\partial_x h_\pm\sigma^*_{xy}|_{h_\pm} + \sigma_{yy}|_{h_\pm} = 0\,. \tag{2.31b}$$

Analogue to the fiber model, the variables are expanded in the order parameter β^2 and, (2.30a) and (2.31a) reveal that the axial velocity does not change across the thickness at leading order, i.e. $\mathcal{O}(\beta^0)$,

$$\sigma^{*(0)}_{xy} = \partial_y u^{(0)} = 0\,. \tag{2.32}$$

Depth-averaging (2.29), utilizing (2.26), leads to the conservation equation

$$\partial_t h + \partial_x\left(hu^{(0)}\right) = 0\,. \tag{2.33}$$

Equations (2.30b) and (2.31b) directly yield $\sigma^{(0)}_{yy} = 0$, showing that the leading order pressure $p^{(0)}$ is identical to $\tau^{(0)}_{yy} = -2\partial_x u^{(0)}$. Consequently, the normal stress difference becomes

$$\sigma^{(0)}_{xx} = \nu^{(0)} = \tau^{(0)}_{xx} - \tau^{(0)}_{yy} = 4\partial_x u^{(0)}\,, \tag{2.34}$$

where the Trouton ratio of 4 indicates purely *planar deformation*, which means that the external, deforming stresses primarily act along one axis and are additionally

[3] A more rigorous way to handle the Laplace pressure, similar to the fiber drawing model, would result in a negligible axial curvature at leading order.

chosen such that the width is kept constant. As the film width is assumed to be infinite (assumption vi), this premise is implicitly fulfilled.[4] Integrating the β^2-order of (2.30a), using (2.31a), finally leads to

$$Re\, h\left(\partial_t u^{(0)} + u^{(0)}\partial_x u^{(0)}\right) = \partial_x\left(h\nu^{(0)}\right) + \frac{Re}{Fr}h\,. \tag{2.35}$$

The system of Eqs. (2.33), (2.34) and (2.35) determines the evolution of thickness and axial velocity of the film casting process at leading order, and the superscript "(0)" will again be omitted from now on. The boundary conditions are, analogue to (2.25),

$$h(x=0,t) = u(x=0,t) = 1\,, \tag{2.36a}$$

$$u(x=1,t) = Dr\,. \tag{2.36b}$$

Comparing the fiber model, i.e., (2.16), (2.18), (2.24) and (2.25), with the film model, i.e., (2.33), (2.34), (2.35) and (2.36), it can be seen that they are almost identical in the case of negligible surface tension, i.e., for $1/Ca^* \to 0$. The only difference is the Trouton ratio,[5] which is constant in both cases, though, and could be shifted into the Reynolds number. Therefore, the dynamics of slender Newtonian fiber drawing, neglecting surface tension, and infinite-width Newtonian film casting are equivalent. This would not be the case anymore for non-Newtonian fluids for which the equations for the normal stress difference (2.18) and (2.34) will differ more than by a numerical prefactor, as shown in Chap. 9. Similarly, the equation for the conservation of energy and its one-dimensional version will also differ between the film and the fiber configurations, treated therefore separately in Chaps. 10 and 11.

---

[4] This also shows that the terms "infinite width" film casting and "constant width" film casting can be used interchangeably here.

[5] Back to dimensional equations, the Trouton ratio would appear as a prefactor of the dynamic viscosity μ, such as one can speak of and effective viscosity 3μ and 4μ for, respectively, the fiber and the film configurations, also referred to as the *extensional viscosity* in literature.

Chapter 3
Linear Stability

Abstract In this chapter, we introduce the linear stability analysis for the particular configuration of infinite-width film casting and in the purely viscous case. Note that this configuration is identical to the fiber drawing one with neglecting surface tension, as demonstrated in Chap. 2. Next we present an equivalent formulation that includes the normal stress difference as additional dependent variable. This alternative formulation offers a more general framework for the extension of the purely viscous case to incorporate all additional effects such as gravity, inertia, surface tension (for fiber drawing), neck-in (for film casting), viscoelasticity, and non-isothermal effects that will be considered in the rest of the book. Two methods to solve the eigenvalue problems resulting from the linear stability analysis are then presented, namely the Chebyshev colocation and the continuation methods, discussing the advantage and inconvenience for both. The last section gives more insights on the physical mechanism of the draw resonance instability and presents alternative criteria for the critical draw ratio based on kinetic wave theory.

Model equations derived in Chap. 2 are coupled nonlinear partial differential equations, and there are several ways to analyze the dynamics of such systems. One possibility would be a direct numerical solution, which is mostly done with finite element methods. While there exists a multitude of algorithms for such a procedure and sufficient computational power nowadays, other analysis methods may still be the better choice, depending on the desired information. In the scope of draw resonance, the focus lies primarily on the detection of the critical draw ratio, i.e., the onset of the draw resonance instability. Actually, the instability leading to rupture or undesirable defects on drawn fibers and films, it is crucial to determine the critical point below which the system remains stable. A full numerical analysis requires solving for transient solutions of the equations for various values of the draw ratio, followed by a post-processing analysis to decide at which draw ratio the instability sets in (see e.g., Fig. 1.5). Because this procedure is time-consuming and of rather low accuracy, thus preventing to explore the critical conditions in the whole space of parameters, we will rather take benefit of the linear properties of the system behavior at criticality and perform a linear stability analysis to identify the critical point.

© Springer Nature Switzerland AG 2025

M. Bechert and B. Scheid, *Draw Resonance*, Fluid Mechanics and Its Applications 141,
https://doi.org/10.1007/978-3-031-85568-9_3

3.1 Pure Viscous Case

Good introductions into the general topic of hydrodynamic instability can be found in the seminal textbook of Chandrasekhar (1961) or in more recent ones, such as for instance Drazin and Reid (2004), Manneville (2010), Charru (2011) or Gallaire and Brun Gallaire and Brun (2017). A general overview of stability theory is provided by the textbook of Strogatz (2000). Following these lines, the linear stability analysis is first applied here to draw resonance on the basis of the one-dimensional film casting model in the absence of inertia and gravity. The system of nonlinear partial differential equations (PDE's) to be solved in this context is obtained by writing the conservation equation (2.33) and by integrating the depth-averaged momentum equation (2.35) with $Re = 0$, and in which (2.34) has been used,

$$\partial_t h + \partial_x (hu) = 0 \,, \tag{3.1a}$$

$$h \partial_x u = f \,, \tag{3.1b}$$

where $f = f(t)$ is an unknown integration constant proportional to the film tension[1] $f^* = h\sigma_{xx}$ by a factor 4 corresponding to the Trouton ratio. Notice that this tension is space-independent in the purely viscous case. The system above should be solved with the following boundary conditions (2.36a), reproduced here for the sake of completeness,

$$h(0, t) = 1 \,, \tag{3.2a}$$

$$u(0, t) = 1 \,, \tag{3.2b}$$

$$u(1, t) = Dr \,. \tag{3.2c}$$

Note that only the two first boundary conditions are needed to solve (3.1) as an *initial value problem*. The third condition is nevertheless needed to determine the film tension $4f$, providing a direct relationship with the draw ratio Dr.

Because we are interested in the stability of stationary solutions and more particularly the critical point for their instability, we first seek for these stationary solutions, also referred to as *base state* solutions. It consists in imposing in (3.1) time-independent variables as follow,

$$h(x, t) = h_s(x) \,, \tag{3.3a}$$

$$u(x, t) = u_s(x) \,, \tag{3.3b}$$

$$f(t) = f_s \,, \tag{3.3c}$$

which gives the following equations, from (3.1), also referred to as the *fixed-point condition* in the context of *dynamical system theory*,

[1] The film tension refers to the axial force per unit length in the transverse direction of the film, which is equivalent to the force rescaled by the constant width. Force and tension are thus equivalent and therefore used indifferently in the case of infinite-width film casting.

$$(h_s u_s)' = 0\,, \tag{3.4a}$$
$$h_s u_s' = f_s\,, \tag{3.4b}$$

where the prime denotes the derivative with respect to x. The corresponding boundary conditions from (3.2) merely become

$$h_s(0) = 1\,, \tag{3.5a}$$
$$u_s(0) = 1\,, \tag{3.5b}$$
$$u_s(1) = Dr\,. \tag{3.5c}$$

The system of equations (3.4) can be integrated once, using the inlet boundary conditions at $x = 0$ to yield

$$h_s u_s = 1\,, \tag{3.6a}$$
$$u_s = \mathrm{e}^{f_s x}\,. \tag{3.6b}$$

Equation (3.6a) indicates a constant flow rate along the film casting direction, as imposed by the continuity equation (2.33). Therefore, it demonstrates the thinning effect of the casting process, namely that the thickness decreases as the velocity increases, while (3.6b) shows that this velocity increase is exponential in space. Using now the outlet boundary condition at $x = 1$ gives $f_s = \ln(Dr)$, which finally leads to the stationary solutions from (3.6),

$$u_s(x) = Dr^x\,, \tag{3.7a}$$
$$h_s(x) = Dr^{-x}\,. \tag{3.7b}$$

As we are interested in the stability of the stationary solutions (3.7), we next look at the behavior of the nonlinear PDE's (3.1) with respect to small perturbations around these stationary solutions. To this aim, we seek for solutions of the form

$$h(x,t) = h_s(x)(1 + \epsilon\,\tilde{h}(x,t))\,, \tag{3.8a}$$
$$u(x,t) = u_s(x)(1 + \epsilon\,\tilde{u}(x,t))\,, \tag{3.8b}$$
$$f(t) = f_s(1 + \epsilon\,\tilde{f}(t))\,, \tag{3.8c}$$

where $\tilde{h}$, $\tilde{u}$ and $\tilde{f}$ are the perturbations normalized by the stationary solutions,[2] and $\epsilon \ll 1$ is a small ordering parameter allowing to linearize the system (3.1) when expanding in power of ϵ and truncating the $\mathcal{O}(\epsilon^2)$ terms coming from nonlinearities. As the zero order is identically satisfied by the fixed-point condition, it brings us to the following *perturbation equations*, together with the boundary conditions:

[2] Note that this normalization is convenient as it reduces the number of terms in the perturbation equations, as compared to those obtained by imposing instead $h(x,t) = h_s(x) + \epsilon\,\tilde{h}(x,t)$, $u(x,t) = u_s(x) + \epsilon\,\tilde{u}(x,t)$ and $f(t) = f_s + \epsilon\,\tilde{f}(t)$

$$h_s \partial_t \tilde{h} + \partial_x (\tilde{h} + \tilde{u}) = 0\,, \tag{3.9a}$$

$$\partial_x \tilde{u} + f_s (\tilde{h} + \tilde{u}) = f_s \tilde{f}\,, \tag{3.9b}$$

$$\tilde{h}(0,t) = 0\,, \quad \tilde{u}(0,t) = 0 \quad \text{and} \quad \tilde{u}(1,t) = 0\,, \tag{3.9c}$$

in which the stationary equations (3.4) have been used. Because the linear system (3.9) is *autonomous* in time, the solutions should have the forms

$$\tilde{h}(x,t) = H(x)\, \mathrm{e}^{\omega t}\,, \tag{3.10a}$$

$$\tilde{u}(x,t) = U(x)\, \mathrm{e}^{\omega t}\,, \tag{3.10b}$$

$$\tilde{f}(t) = F\, \mathrm{e}^{\omega t}\,, \tag{3.10c}$$

where ω is called the *eigenvalue*, which is complex and associated to the complex *eigenfunctions* H and U, while F is a constant that sets the amplitude of the perturbations. Posing $\omega = \omega_R + \mathrm{i}\,\omega_I$, the stability is determined through the sign of ω_R: for $\omega_R < 0$, the initial perturbation decays in time and the system is stable, while for $\omega_R > 0$, the initial perturbation grows in time and the system is unstable. The critical point we are seeking for thus corresponds to the transition with $\omega_R = 0$, also referred to as the *neutral stability mode*. Additionally, for $\omega_I \neq 0$, the system is *oscillatory*, otherwise, for $\omega_I = 0$ the system is *monotonous*.

Introducing (3.10) into (3.9) gives, using $h_s = \mathrm{e}^{-f_s x}$,

$$\mathrm{e}^{-f_s x} \omega H + H' + U' = 0\,, \tag{3.11a}$$

$$U' + f_s (H + U) = f_s F\,, \tag{3.11b}$$

$$H(0) = 0\,, \quad U(0) = 0 \quad \text{and} \quad U(1) = 0\,. \tag{3.11c}$$

Because all boundary conditions are equal to zero, the eigenvalue problem corresponds to an *homogeneous initial-value problem*, which can be determined safe one constant. Therefore, one can arbitrarily set the value of F.[3] By rearranging (3.11b) as

$$H = F - U - \frac{U'}{f_s}\,, \tag{3.12}$$

and deriving (3.11a) once with respect to x, one obtains a second-order ordinary differential equation for the variable U of the form

$$U'' = f_s \omega \mathrm{e}^{-f_s x} \left(F - U - \frac{U'}{f_s} \right)\,, \tag{3.13}$$

to be solved with the following inlet boundary conditions,

---

[3] Setting $F = 1$ could appear as the most general choice but it would more rapidly lead to unrealistic, even negative, values for the thickness (see, e.g., Figs. 3.1 and 3.2). Unless specified otherwise, the choice of $F = 0.1$ will thus be made for the sake of convenience in the following.

$$U(0) = 0 \quad \text{and} \quad U'(0) = F\, f_s\,, \tag{3.14}$$

where the second boundary condition corresponds to (3.11b) evaluated in $x = 0$. Equation (3.13) has an analytical solution that has been formulated by Schultz and Davis (1984) and Renardy (2006), and can easily be obtained using a software for symbolic calculus[4]:

$$U = \frac{1}{f_s} e^{-\frac{\omega}{f_s} - f_s x} \left[\left(f_s\, e^{f_s x} - \omega \right) \left\{ \mathrm{Ei}\left(\frac{\omega}{f_s} \right) - \mathrm{Ei}\left(\frac{\omega}{f_s} e^{-f_s x} \right) \right\} + f_s\, e^{f_s x} \left(e^{\frac{\omega}{f_s}} - e^{\frac{\omega}{f_s} e^{-f_s x}} \right) \right] \tag{3.15}$$

$$\text{with} \quad \mathrm{Ei}(z) = \int_{-\infty}^{z} dt\, \frac{e^t}{t}\,. \tag{3.16}$$

Injecting (3.15) into (3.12) gives

$$H = e^{-\frac{\omega}{f_s}} \left\{ -\mathrm{Ei}\left(\frac{\omega}{f_s} \right) + \mathrm{Ei}\left(\frac{\omega}{f_s} e^{-f_s x} \right) \right\}. \tag{3.17}$$

Finally the use of the outlet boundary condition $U(1) = 0$ provides the characteristic equation

$$\left[\left(f_s\, e^{f_s} - \omega \right) \left\{ \mathrm{Ei}\left(\frac{\omega}{f_s} \right) - \mathrm{Ei}\left(\frac{\omega}{f_s} e^{-f_s} \right) \right\} + f_s\, e^{f_s} \left(e^{\frac{\omega}{f_s}} - e^{\frac{\omega}{f_s} e^{-f_s}} \right) \right] = 0\,, \tag{3.18}$$

which needs to be solved in order to calculate the eigenvalue spectrum in dependence of f_s, or alternatively the draw ratio Dr. In other words, as (3.18) is complex, one has two equations for three unknowns, namely f_s, ω_R and ω_I. Consequently, one parameter should be fixed in order to determine the two other ones. This equation can be solved by numerical means only. Therefore, the solution depends on the initial guess on the parameter values, which need to be in the vicinity of the eigenvalue with largest real part, as this corresponds to the most unstable mode. In particular, the critical draw ratio can be determined by setting $\omega_R = 0$ and applying a numerical root finding algorithm for solving (3.18) with ω_I and f_s as variables. For this critical mode in the purely viscous case, one finds the classical results identical to the one-dimensional fiber (without surface tension) or film model (Pearson and Matovich 1969; Gelder 1971; Yeow 1974):

$$\omega_{I_c} = 14.0111 \quad \text{and} \quad f_{s_c} = 3.00657 \quad \text{or} \quad Dr_c = 20.2179\,. \tag{3.19}$$

The associated stationary and eigenfunction solutions are plotted in Fig. 3.1 at criticality.

[4] Using Mathematica© for instance, the command is `DSolve[{U″[x] == fs ω Exp[-fs x] (F - U[x] - U′[x]/fs), U[0] == 0, U′[0] == F fs}, U[x], x]`.

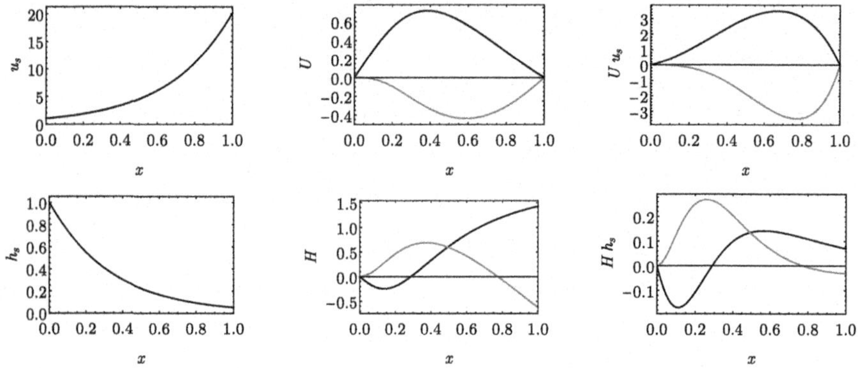

Fig. 3.1 Stationnary solutions u_s and h_s, eigenfunctions U and H, and perturbation amplitudes u_sU and h_sH at criticality, i.e. for $Dr_c = 20.218$, with $F = 1$. Real parts are in black lines and imaginary parts in grey lines

Table 3.1 The six first eigenvalues at criticality, i.e. for $Dr = 20.218$. Note that these eigenvalues should be scaled by the factor $f_s = \mathrm{e}^{Dr}$ to correspond to those reported by Renardy (2006)

n	ω_R	ω_I
1	0	14.011
2	−2.2208	34.435
3	−3.6194	54.856
4	−4.6861	75.168
5	−5.5659	95.401
6	−6.3262	115.58

It can be observed that the perturbations, i.e. the eigenfunctions modulated by the stationary solutions, have their maxima shifted to the outlet for the velocity and to the inlet for the thickness.

The mode (3.19) above corresponding to the vanishing of the highest real part of the eigenvalue, it can be completed by the higher-order modes at criticality, as computed using standard solver for eigenvalue problems (see next section). This is reported in Table 3.1 for the first 6 modes referenced by the index n.

To further illustrate the linear behavior of the system, one can decompose $U = U_R + \mathrm{i}\, U_I$ and $H = H_R + \mathrm{i}\, H_I$ in order to reconstruct the thickness and velocity variables by only taking the real part of the perturbations (3.10), which gives

$$h(x,t) = h_s(x)\left(1 + [H_R(x)\cos(\omega_I t) - H_I(x)\sin(\omega_I t)]\,\mathrm{e}^{\omega_R t}\right), \tag{3.20a}$$

$$u(x,t) = u_s(x)\left(1 + [U_R(x)\cos(\omega_I t) - U_I(x)\sin(\omega_I t)]\,\mathrm{e}^{\omega_R t}\right). \tag{3.20b}$$

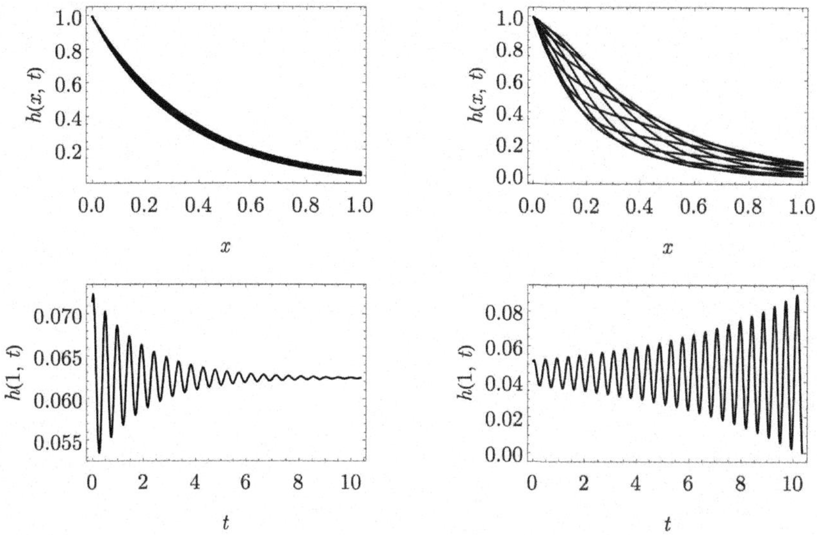

(a) $Dr = 16 < Dr_\mathrm{c}$ (b) $Dr = 22 > Dr_\mathrm{c}$

Fig. 3.2 Linear solutions as reconstructed from (3.20a) for two different draw ratios 16 and 22, with an arbitrarily small perturbation value of $F = 0.1$. The corresponding highest eigenvalues are (**a**) $\omega = -0.49422 + \mathrm{i}\,13.278$ and (**b**) $\omega = 0.1816 + \mathrm{i}\,14.276$. The upper plots show the thickness profile at ten different times regularly spaced within the first period of oscillation for (**a**) and the last period of oscillations for (**b**), while the bottom plots show the time evolution of the thickness at the outlet

These linear solutions are plotted in Fig. 3.2 for two draw ratios, one below and one above the threshold. Thickness profiles are shown (see upper plots) at intervals of $T/10$ where $T = 2\pi/\omega_I$ is the period of oscillation. As inferred from the sign of the real part of the maximum eigenvalue, the perturbation is damped for $Dr = 16 < Dr_\mathrm{c}$ and amplified for $Dr = 22 > Dr_\mathrm{c}$, as illustrated by the time evolution of the thickness at the outlet $h(1, t)$ (see bottom plots). Note that the time evolution for the unstable situation in (b) stops when $h(1, t)$ becomes negative as it would in practice lead to the breakage of the film.[5]

3.2 Linear Stability: General Framework

We propose in this section an alternative, though equivalent, analysis than the one for the purely viscous case presented in the previous section. This consists in keeping the normal-stress difference ν as an additional dependent variable, namely

[5] Note that with a larger perturbation amplitude, e.g., $F = 1$, $h(1, t)$ would have become negative immediately for (a) and after only few periods for (b).

$$\partial_t h + \partial_x (hu) = 0\,, \tag{3.21a}$$
$$\partial_x (h\nu) = 0\,, \tag{3.21b}$$
$$4\partial_x u = \nu\,. \tag{3.21c}$$

Note the boundary conditions (3.2) remain unchanged. Even though ν can be eliminated in the pure viscous case, it brings interesting information for further extensions. The relationship between the normal-stress difference and the film tension $4f$ is

$$4f = h\nu\,. \tag{3.22}$$

As the number of eigenvalues in an eigenvalue problem is in principle infinite, instead of writing the perturbation equations as in the previous section, let us expand the variables as follows,

$$h(x,t) = h_s(x)\left[1 + \sum_{n=1}^{\infty}\left(H_n(x)\,\mathrm{e}^{\omega_n t} + H_n^*(x)\,\mathrm{e}^{\omega_n^* t}\right)\right], \tag{3.23a}$$
$$u(x,t) = u_s(x)\left[1 + \sum_{n=1}^{\infty}\left(U_n(x)\,\mathrm{e}^{\omega_n t} + U_n^*(x)\,\mathrm{e}^{\omega_n^* t}\right)\right], \tag{3.23b}$$
$$\nu(x,t) = \nu_s(x)\left[1 + \sum_{n=1}^{\infty}\left(N_n(x)\,\mathrm{e}^{\omega_n t} + N_n^*(x)\,\mathrm{e}^{\omega_n^* t}\right)\right], \tag{3.23c}$$

where ν_s is the stationary normal-stress difference, N is the associated complex eigenfunction, n is the mode index and the asterisk denotes the complex conjugate. As seen in the previous section, the crucial step, after plugging (3.23) into (3.21), is to limit the perturbations to small amplitudes, i.e., $|U_n|$, $|H_n|$, $|N_n| \ll 1$, and to linearize the equations. Due to the linearization, the complex conjugated part in (3.23) can be calculated separately. Moreover, as each perturbation mode forms a separate system of equations, the mode index n is dropped from now on for the sake of simplicity, such as the variables are expanded exactly like in the previous section, i.e.

$$h(x,t) = h_s(x)\left(1 + H(x)\,\mathrm{e}^{\omega t}\right), \tag{3.24a}$$
$$u(x,t) = u_s(x)\left(1 + U(x)\,\mathrm{e}^{\omega t}\right), \tag{3.24b}$$
$$\nu(x,t) = \nu_s(x)\left(1 + N(x)\,\mathrm{e}^{\omega t}\right), \tag{3.24c}$$

together with the note that there is an infinite number of eigenvalues and associated eigenfunctions.

Following the same steps as in the previous section, the stationary solutions of (3.21) are

$$h_s(x) = Dr^{-x}\,, \tag{3.25a}$$
$$u_s(x) = Dr^{x}\,, \tag{3.25b}$$
$$\nu_s(x) = 4\ln(Dr)Dr^{x}\,. \tag{3.25c}$$

And the linearized perturbation equations are

$$H' + U' = -\omega h_s H \,, \tag{3.26a}$$

$$H' + N' = 0 \,, \tag{3.26b}$$

$$\nu_s N - 4\left(u_s' U + u_s U'\right) = 0 \,, \tag{3.26c}$$

closed with the boundary conditions (3.11c). As seen earlier, an analytical solution exists for this eigenvalue problem, but in case it does not, which will be the case with any of the additional effects that will be considered in the subsequent chapters, such an eigenvalue problem needs to be solved numerically following one of the two methods outlined below.

3.2.1 Chebyshev Collocation Method

The first method consists of a discretization of the system, which makes it possible to write the former continuous differential equations (3.26) in a discrete form,

$$\mathbf{A} \cdot \mathbf{z} = \omega \mathbf{B} \cdot \mathbf{z} \,, \tag{3.27}$$

where $\mathbf{z}$ is a $3(J+1)$-dimensional vector containing the discretized spatial perturbations, with J being the order of discretization, and $\mathbf{A}$ and $\mathbf{B}$ are both $3(J+1) \times 3(J+1)$ matrices, which are specified by the perturbation equations and the boundary conditions. The eigenvalues and eigenfunctions of such a system can be solved with the help of well-established numerical methods. The accuracy and robustness of the solution, however, depends strongly on the way the system is discretized. In this book Chebyshev collocation will be used, following Hagen and Langwallner (2006), as this method appears to converge already for $J \gtrsim 100$. In general, $\mathbf{A}$ and $\mathbf{B}$ depend on the stationary solutions. As in most cases the steady state equations cannot be solved analytically, a numerical shooting method is applied here to first determine the steady state. The entire procedure is explained in more detail in Appendix A.

Figure 3.3 shows the eigenvalues of (3.26) in the complex plane for draw ratios of 16 and 22. The complex eigenvalues appear always in conjugated pairs. For stable process conditions ($Dr = 16$), all eigenvalues have negative real parts. If the draw ratio becomes larger than a critical value Dr_c, e.g., $Dr = 22$, at least one pair of complex eigenvalues has a positive real part and the system becomes unstable, as it can be seen from (3.24) that the real part of the eigenvalues determines whether an initial perturbation increases or decreases with time. An instability occurring if a pair of complex eigenvalues crosses the imaginary axis is called a *Hopf bifurcation*.

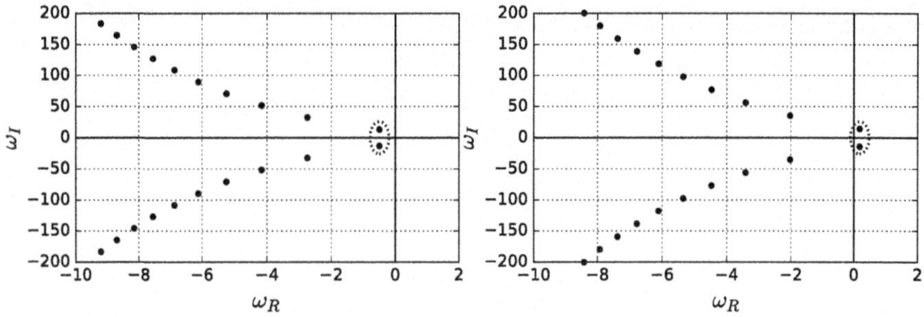

Fig. 3.3 Eigenvalue spectrum of the linearized equations of the purely viscous, infinite-width film casting model for draw ratios of 16 and 22. For stable conditions, all eigenvalues have real parts smaller than zero. The instability sets in if at least one pair of complex eigenvalues (*circled*) crosses the imaginary axis such that the real part becomes positive

With increasing draw ratio $Dr \geq Dr_c$, the amplitude of the oscillations increases continuously from zero (see Sect. 1.2, Fig. 1.6; Yun et al. 2008; Burghelea et al. 2012) and therefore draw resonance is a *supercritical Hopf bifurcation*.[6]

3.2.2 *Continuation Method*

The second method to solve the eigenvalue equations consists in solving the boundary value problem with ω as a free parameter. In this case, an additional boundary condition is needed. As the eigenvalue equations are homogeneous, the amplitude of the perturbations can be scaled arbitrarily. In most cases throughout this book, the perturbation of the film tension $h\nu$ is fixed at the outlet to 0.1, i.e.,

$$H(1) + N(1) = 0.1 \; , \tag{3.28}$$

as the force oscillations and the relative phase to other oscillation modes appears to be crucial for investigating the physical mechanism underlying draw resonance (see also Sect. 3.3), and fixing the force amplitude at the outlet may simplify the analysis. Nevertheless, this is an arbitrary choice and does not affect the dynamics in any way. Computing the eigenvalue spectrum implies to find the values of ω for which the solutions fulfill the boundary conditions (3.11c) and (3.28), which is in general a challenging task. However, as described above, it is not the complete eigenvalue spectrum that is needed to determine the stability, but merely the eigenvalue pair with the largest real part. This fact can be exploited by numerical continuation. The

[6] The information about the growing behavior of the oscillation amplitude cannot be gained by linear stability analysis, likewise for a *subcritical instability*, which sets in for $Dr \leq Dr_c$ with a finite amplitude perturbation

basic assumption of this method is that all solutions lie on a continuous branch in the parameter space. Then it is possible to track the solutions, i.e., the steady states and the perturbations, by continuous variation of the control parameters, provided one set of solutions is known for a particular parameter combination and can be used as a starting point. A good choice for such a starting point is the analytical solution of the purely viscous linear stability problem, namely (3.15) and (3.17), using the highest eigenvalue as calculated from the characteristic equation (3.18). This is the approach adopted in this book and detailed below.

In general, starting from the initially known solution, the draw ratio is varied as primary continuation parameter together with ω_R and ω_I as secondary continuation parameters, until ω_R becomes zero, which defines the critical point, as described by (3.19). The value of ω_I then indicates the oscillation frequency at criticality. The steady states and complex perturbations are plotted in Fig. 3.4, as in previous section, but complemented with the solution for the normal stress-difference, and with the steady states in log-log plots.

Afterwards, possible additional control parameters induced by more complex models can be varied by keeping $\omega_R = 0$ and instead using the control parameter as third continuation parameter. Such a three-parameter numerical continuation can be performed very efficiently with the software AUTO-07P (Doedel et al. 2012). as explicated in a tutorial given in Appendix B. Written in Fortran, AUTO-07P is basically controlled with the help of a code file containing the system of differential equations together with the boundary conditions and the initial conditions of the starting solution, and a configuration file, within which the continuation method is specified further, e.g., by setting the size of one continuation step or the level of accuracy. As the computed critical parameter sets are obtained by an iterative

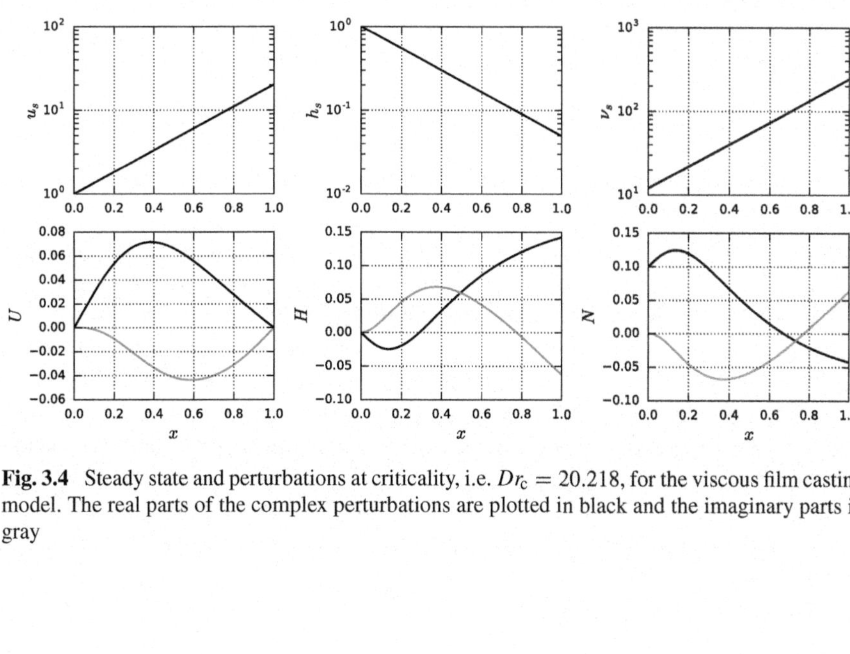

Fig. 3.4 Steady state and perturbations at criticality, i.e. $Dr_c = 20.218$, for the viscous film casting model. The real parts of the complex perturbations are plotted in black and the imaginary parts in gray

approximation to a continuous curve instead of a point-wise calculation based on a fixed mesh, the neutral stability curves, i.e., the curves corresponding to critical conditions, are visualized by continuous lines.

3.2.3 *Additional Remarks on the Linear Stability Analysis*

Despite the high efficiency and flexibility of AUTO-07P, this method fails if another perturbation mode than the originally tracked one becomes the most unstable during the continuation process. In this case, the full eigenvalue spectrum will be calculated using the discretization method presented in Sect. 3.2.2 and the critical parameter set is determined by a stepwise variation of the draw ratio, until the largest real part of the eigenvalues becomes zero. This procedure becomes necessary in parts of Sect. 8.1.

Linear stability analysis is a powerful tool for determining the critical draw ratio with respect to other control parameters, but it is only valid in the vicinity of criticality. For further information, e.g., the oscillation amplitude evolution together with the sub/supercritical nature of the instability, other methods like *weakly nonlinear analysis* or full numerical computations (as illustrated in Sect. 1.2) need to be applied but this will not be treated in this book.

It should finally be noted that the ansatz (3.23) neglects the perturbations in the film width direction in the case of film casting processes. One reason for this is that it is the goal to keep the analysis as simple as possible, and accounting for transversal perturbation waves would introduce a further parameter, the wave number. In particular cases it has been shown that the critical draw ratio becomes minimal if the wave number is zero (Yeow 1974; Ahmed and Khayat 2012), therefore the most unstable scenario is already covered with ansatz (3.23). Nevertheless, this may not be true in general (e.g., Aird and Yeow (1983)), as it depends on the particular model assumptions.

3.3 Physical Mechanism and Alternative Stability Criteria

In draw resonance, and because of the nature of the equations, none of the dependent variables are in phase with each other, at any position from the inlet to the take-up, which challenges both intuition and understanding of the basic mechanisms underlying the instability. In isothermal conditions and for Newtonian fluids, Hyun (1978) proposed an approximate criterion for the occurrence of draw resonance based on the fact that the mass of fluid that enters the system at a constant rate, i.e., constant throughput, must leave the system in a fixed amount in every cycle of draw resonance. Hyun (1978) established therefore that for the draw resonance instability to occur, a full cycle of disturbance waves, also called throughput wave, whose traveling time is noted t_L, should travel through the system in a time ($2t_L$) smaller than the residence time, defined as

$$\tau_L = \int_0^1 \frac{dx}{u_s} \,. \tag{3.29}$$

Nevertheless, Hyun's criterion predicts analytically a critical draw ratio of 19.774 slightly smaller than the correct numerical value of 20.218, as pointed out by Denn (1980b) (see also the rebutal by Hyun (1980)). Anyway, Hyun's criterion has the merit to rely on intuition, the mechanism of which is reviewed below, in light with the refinement of the criterion by Kim et al. (1996) using kinematic wave theory.

3.3.1 Kim's and Jung's Criteria

Alternatively to the linear stability analysis, insights to the physical mechanism causing the instability can be gained by analyzing the full numerical solutions of (3.1), like those shown in Fig. 1.5. More specifically, by observing the time-dependent evolution of the variables at criticality and above for different axial positions, Kim et al. (1996) proposed a draw resonance mechanism and a stability criterion, which are both based on the traveling times of various perturbation waves. This approach is shortly reviewed here on the basis of the purely viscous one-dimensional film model, in line with the previous linear stability analysis.

The dimensionless force f^* corresponding to the normal stress difference can be calculated with the help of (2.34) as

$$f^* = \int_{-h/2}^{h/2} \nu \, dy = 4h\partial_x u \,. \tag{3.30}$$

As already mentioned, the momentum equation (3.1b) shows that $f = f^*/4$ does not change along the x-axis, which means in particular that the force acting at the outlet is identical to the force acting at the inlet. It thus implies that if the force is fluctuating at the outlet, it should instantaneously lead to a change of the thickness and the velocity at the inlet, i.e. $x = 0$, as (3.1) can be transformed into

$$\partial_x h|_0 = -f \,, \tag{3.31}$$

where the boundary conditions (3.2) have been used.[7] Equation (3.31) shows that a temporal maximum of the pulling force leads to a minimum of the thickness directly after the inlet. This minimum thickness then travels towards the outlet. Shortly before it reaches the outlet, the flow perturbations lead to a minimum pulling force, as the numerical investigations of Kim et al. (1996) reveal. A full cycle of perturbation waves thus includes a 'positive' wave for which $\mathrm{d}f/\mathrm{d}t > 0$ and a 'negative' wave for which $\mathrm{d}f/\mathrm{d}t < 0$. Note that the more intuitive picture that a minimum thickness

[7] With $h = u = 1$ in $x = 0$, (3.1a) simplifies into $\partial_x u|_0 = -\partial_x h|_0$, which have then been substituted into (3.1b).

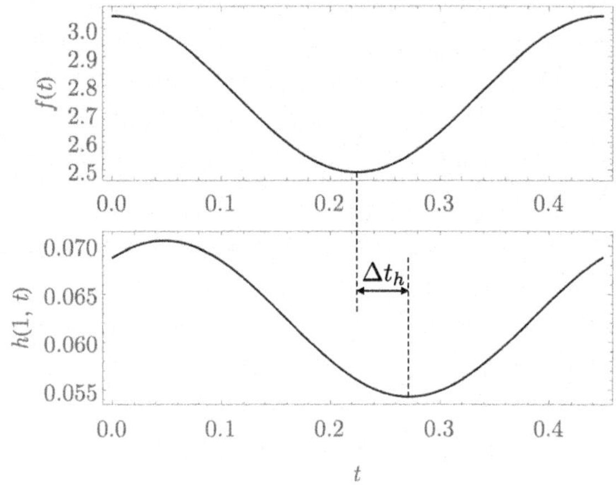

Fig. 3.5 Time delay Δt_h between the tension and the thickness at take-up, the former leading the latter. It shows the first oscillation for $Dr = 16$ in Fig. 3.2a

leads to a minimum force may help imagination but is technically inaccurate. Indeed, as also stated by Kim et al. (1996), (3.30) shows that f and h cannot be in phase unless $\partial_x u$ is in phase too. Therefore we denote $\Delta t_h > 0$ the time delay between the thickness and the force oscillations at the outlet, assuming the force leads the thickness, as illustrated in Fig. 3.5.

A complete cycle of draw resonance is illustrated in Fig. 3.6. The sequence for the first half of the cycle (labelled from 1 to 4) is as follows: the temporal minimum of the force instantaneously induces, following (3.31), a minimum slope of the thickness at the inlet (1), $\partial_x h|_0$, and consequently a maximum thickness in the vicinity of the inlet (2), say at $x = 0^+$. This maximum thickness travels towards the outlet during the time t_L (3), slightly preceded by a maximum of the force, Δt_h earlier (4), and inducing the second sequence of the cycle (labelled from 5 to 8) involving anti-symmetrically a wave of minimum thickness.

Using the force fluctuations as basis of the instability mechanism also explains the findings of Schultz and Davis (1984), who investigated the influence of different boundary conditions at the outlet. Instead of fixing the outlet speed, they formulated a mixed boundary condition, which can be continuously tuned between fixed outlet speed and fixed outlet force. Their results show that the critical draw ratio increases if the boundary condition is shifted towards fixed outlet force and finally becomes infinite in the case of a completely fixed force, meaning an unconditionally stable system. The terminology "unconditionally stable" is to be understood in terms of infinitesimal perturbations and does not make any statements on the finite amplitude stability. Interestingly enough Renardy (2006) extended the study on different boundary conditions of fiber drawing and showed that a prescribed outlet fiber diameter leads to "unconditionally unstable" dynamics.

In any case, these studies and the arguments above demonstrate the crucial role of the tension in providing the feedback mechanism for sustaining draw resonance. Indeed, as rationalized by Kim et al. (1996), this feedback mechanism can only

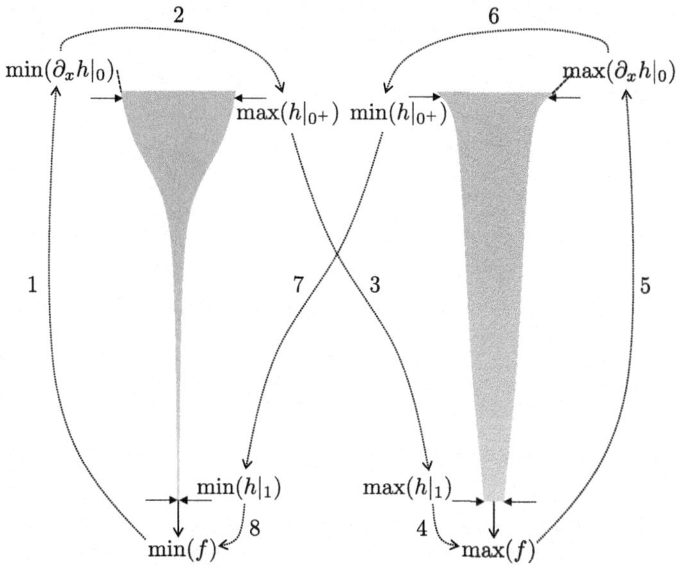

Fig. 3.6 Sketch of a cycle of draw resonance. Steps 1, 2, 5 and 6 are instantaneous, steps 3 and 7 have a duration t_L of a unity-throughput wave, and steps 4 and 8 have a negative duration of $-\Delta t_h$, given that the tension f precedes the thickness $h|_1$, as shown in Fig. 3.5. Note the two snapshots used for this sketch have been extracted form Fig. 1.7 that also shows intermediate snapshots during the traveling time of maximum/minimum thickness waves

occur if the oscillation period of draw resonance T and the traveling times of the perturbation waves, e.g., the maximum-thickness wave or the unity-throughput wave defined by $q \equiv uh = 1$, with throughput or flow rate q, fulfill the following criterion,

$$(t_L)_1 + (t_L)_2 + \frac{T}{2} \leq (\vartheta_L)_1 + (\vartheta_L)_2 \,, \tag{3.32}$$

where $(\vartheta_L)_1$ and $(\vartheta_L)_2$ are the traveling times of the maximum- and minimum-thickness waves and $(t_L)_1$ and $(t_L)_2$ are the corresponding traveling times of a unity-throughput wave, as illustrated in Fig. 3.7.

As the oscillation is sinusoidal at $Dr = Dr_c$, $(t_L)_1 = (t_L)_2$ and $(\vartheta_L)_1 = (\vartheta_L)_2$ in this case, and (3.32) can be reduced to

$$2t_L + \frac{T}{2} \leq 2\vartheta_L \qquad \text{for} \qquad Dr \geq Dr_c \,. \tag{3.33}$$

This criterion will be referred to as *Kim's criterion* in the rest of this book. As noted by Jung et al. (1999) from observations of the time-dependent simulation results, the equivalence $\vartheta_L = \frac{T}{2} + \Delta t_h$ is verified such as (3.33) can be rewritten as

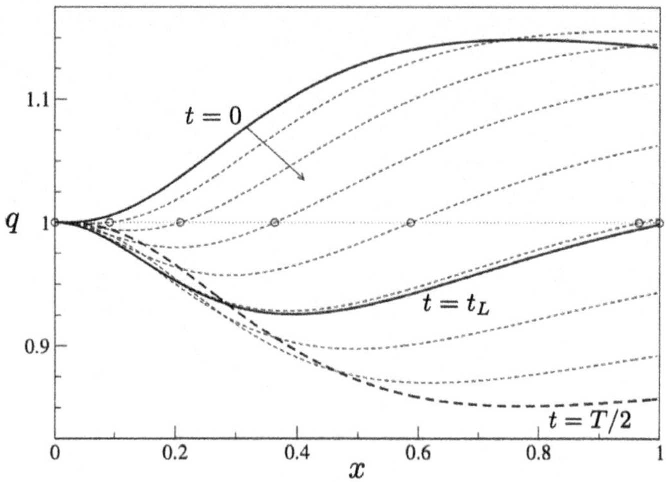

Fig. 3.7 Time evolution of the flow rate q at criticality ($Dr_c = 20.218$) as reconstructed from solutions of the linear stability analysis: the wave starts arbitrarily at $t = 0$ and is plotted at intervals of $T/16$ (dashed lines) over half a period $T/2$, with $T = 0.448$ (the arrow indicates the direction of time). Circles represent the unity-throughput wave for $q = 1$ that starts at $x = 0$ and $t = 0$ and ends at $x = 1$ after a time $t_L = 0.142$. Because of the fixed boundary conditions at inlet, the unity-throughput wave is the only perturbation wave that can travel the entire domain

$$t_L \leq \frac{T}{4} + \Delta t_h \qquad \text{for} \qquad Dr \geq Dr_c \,, \tag{3.34}$$

which as the advantage to be evaluated with quantities that can easily be extracted from the linear stability analysis, as demonstrated in the two next sections.

Because we focus on the linear response of the system, the picture in Fig. 3.7 for a 'negative' unity-throughput wave is symmetric for the second half of the cycle: at time $t = T/2$, when the tension has reached its minimum, the 'positive' unity-throughput wave will start to travel during the same time t_L. It clearly appears from Fig. 3.7 that the only constant-flow-rate wave that travels over the entire domain is the one corresponding to $q = 1$ (see Sect. 3.3.2 for the evaluation of t_L). It also shows that the flow rate averaged over the domain will be as deficient as in excess, compared to the unity base-state flow rate, within one period of draw resonance. In fact, the perturbation equation (3.11b) integrated along the film, using (3.11c), yields

$$\int_0^1 Q \, dx = F \,, \tag{3.35}$$

where $Q = U + H$ is the perturbation of the flow rate. This relation shows that the domain-averaged flow-rate perturbation evolves exactly in phase with the perturbation of the tension.

In an attempt to reconciling Kim's with Hyun's criteria, Jung et al. (1999) derived an approximation of (3.33) involving the fluid residence time (3.29) as well as the time delay, which reads

$$2t_L + \Delta t_h \leq \tau_L \qquad \text{for} \qquad Dr \geq Dr_c \,, \tag{3.36}$$

and which reduces to Hyun's criterion for $\Delta t_h = 0$. This approximative criterion (3.36) will be referred to as *Jung's criterion* in the rest of this book and will be shown to be valid as compared to *Kim's criterion* in the isothermal case, whatever the other effects included, such as gravity and inertia (Chap. 4), surface tension for fiber drawing (Chap. 5) or neck-in for film casting (Chap. 6) and viscoelasticity (Chap. 7). However, *Jung's criterion* fails in the nonisothermal case, as it will be shown in Chap. 10, for which an additional time shift needs to be added on the right-hand side of (3.36).

For the criteria (3.33) and (3.36) to be tested for applicability in the scope of the extended models in subsequent chapters, the relevant quantities will be calculated as follows. The fluid residence time τ_L is obtained from the numerical integration of the inverse of the steady velocity (3.29). The traveling times of maximum- and minimum-thickness waves, ϑ_L, should be determined by tracking an extremal point of a perturbation wave along the x-axis, followed by calculating the difference between the time values at inlet and outlet. The unity-throughput wave traveling time and the time shift Δt_h are derived in the two next sections, combining with results from the linear stability analysis (Sect. 3.1).

3.3.2 Unity-Throughput Time t_L

The mass conservation equation (3.1a) can be written in the form of a kinematic wave equation for a fluid element h, i.e.

$$\partial_t h + u \partial_x h = -h \partial_x u \,, \tag{3.37}$$

where u can be seen as the phase speed of the wave and the right-hand side represents the global attenuation due to the effective tension (since $h\partial_x u = f$ from (3.1b)). Likewise and as in Kim et al. (1996), we can write the kinematic equation for a throughput wave for the quantity $q = uh$ as follows,

$$\partial_t q + c\, \partial_x q = 0 \,, \tag{3.38}$$

where $c = c(x)$ is the phase speed of such a wave. Linearizing (3.38) with $q = 1 + Q\, \mathrm{e}^{\omega t}$ leads to

$$Q' = -\frac{\omega}{c}\, Q \,, \tag{3.39}$$

which can be identified with (3.11a), given that $Q = H + U$, to yield

$$c = \frac{Q}{h_s H} \,. \tag{3.40}$$

Now the only constant-flow-rate wave that travels the entire system is the one corresponding to $q = 1$, referred to as the *unity-throughput wave* (Kim et al. 1996). The time this wave will travel the system is therefore the *unity-throughput time*, obtained as

$$t_L \equiv \int_0^1 \left(\frac{1}{c}\right)_R dx = \int_0^1 h_s \frac{(H_R Q_R + H_I Q_I)}{(Q_R^2 + Q_I^2)} dx \,, \tag{3.41}$$

where we recall that the subscripts R and I refer to real and imaginary parts of complex quantities.

3.3.3 Time Delay Δt_h and General Phase Shifts

In draw resonance, because of the boundary conditions, the tension and the thickness variations are exactly in phase opposition at the inlet, as shown by (3.31), while it is not the case at take-up. Indeed, unless $\partial_x u|_1$ is in phase with $h|_1$, and based on (3.1b), there is no reason for $h|_1$ to be exactly in phase with f. On the basis that f leads h as already assumed in Fig. 3.5, the phase shift between these two quantities can be calculated by applying a trigonometric factorization as follows,

$$H_R \cos(\omega_I t) - H_I \sin(\omega_I t) = H_R \frac{\cos(\omega_I t - \phi)}{\cos(\phi)} \quad \text{with} \quad \tan\phi = -\frac{H_I}{H_R} \,, \tag{3.42}$$

such that the time delay at take-up between $h|_1$ (also $\propto q|_1$) and f has the form

$$\Delta t_h = -\frac{1}{\omega_I} \arctan\left(\frac{H_I(1)}{H_R(1)}\right) . \tag{3.43}$$

Likewise, the phase shift of any averaged quantity $\langle X \rangle$ is generally given by

$$\varphi_{\langle X \rangle} = -\frac{1}{\omega_I} \arctan\left(\frac{\langle X_s X_I \rangle}{\langle X_s X_R \rangle}\right) , \tag{3.44}$$

where X_s is the stationary quantity modulating the corresponding perturbation. In other word, the phase shift is equal to the complex angle of $\langle X \rangle|_1$, as the perturbation of the force is fixed (hence real) at the outlet according to the boundary condition (3.28).

Chapter 4
Influence of Gravity and Inertia

Abstract The purpose of this chapter is to highlight the influences of gravity and inertia on the critical draw ratio. Each effect is treated separately in order to isolate the associated dynamics. Besides the correlations between the critical draw ratio and the particular control parameters, the focus lies here on the physical mechanism underlying the change in stability behavior. Moreover, the alternative stability criteria of Kim and Jung, (3.33) and (3.36), respectively, will be tested for applicability. Simplified versions of the *film model* derived in Sect. 2.2 will be used to study the influence of inertia and gravity. Note that the results will be equivalent for the *fiber model* with neglecting surface tension by simply replacing the thickness h with the cross-section a and the constant Trouton ratio of 4 by 3. The latter can be achieved as well by multiplication of Re or, respectively, Re/Fr with $3/4$, which can be directly seen in the momentum equations.

Hagen and Langwallner (2006) presented a mathematically well-founded numerical analysis with focus on the influence of inertia on the stability of Newtonian fibers. The linear stability analysis reveals a strong increase of the critical draw ratio with increasing Reynolds number, indicating that draw resonance is almost completely suppressed for values of the Reynolds number[1] above 0.1. Numerical solutions of the full nonlinear equations confirmed the linear stability results. German et al. (2006) studied the combined influence of gravity and inertia on the stability behavior. The comprehensive analysis includes methods of linear stability as well as full numerical solutions and the dependencies of the critical draw ratio, the oscillation frequency, the growth rate and fiber rupture on inertia and gravity effects are presented. In general, inertia is said to have a dominant effect on the stability compared to gravity. Reynolds numbers are considered up to a value of 0.2 and the ratio of Reynolds to Froude numbers is limited to 22.

[1] Comparing the absolute values of the Reynolds number in different works may be confusing, as some authors include the constant Newtonian Trouton ratio in the definition of the Reynolds number, and some do not.

© Springer Nature Switzerland AG 2025
M. Bechert and B. Scheid, *Draw Resonance*, Fluid Mechanics and Its Applications 141,
https://doi.org/10.1007/978-3-031-85568-9_4

4.1 Influence of Gravity on Draw Resonance

The *viscous–gravity model* for infinite-width films is governed by the continuity equation (2.33), the constitutive equation (2.34) and the momentum equation (2.35) without the inertial term, i.e.,

$$\partial_t h + \partial_x (hu) = 0 \,, \tag{4.1a}$$

$$\nu - 4\partial_x u = 0 \,, \tag{4.1b}$$

$$\partial_x (h\nu) + \frac{Re}{Fr} h = 0 \,. \tag{4.1c}$$

The steady state equations can be written as

$$h_s' = -\frac{h_s \nu_s}{4u_s} \,, \tag{4.2a}$$

$$u_s' = \frac{\nu_s}{4} \,, \tag{4.2b}$$

$$\nu_s' = \frac{\nu_s^2}{4u_s} - \frac{Re}{Fr} \,, \tag{4.2c}$$

and the linearized perturbation equations resulting from the ansatz (3.24) read

$$H' = -\left(\frac{\omega}{u_s} H + U'\right) \,, \tag{4.3a}$$

$$U' = \frac{\nu_s}{4u_s} (N - U) \,, \tag{4.3b}$$

$$N' = -H' + \frac{Re}{Fr} \frac{N}{\nu_s} \,. \tag{4.3c}$$

The system of Eqs. (4.2) and (4.3) are solved by continuation with the boundary conditions (3.5), (3.11c) and (3.28) following the procedure outlined in Sect. 3.2.2. As depicted in Fig. 4.1a, increasing Re/Fr, and thus the effect of gravity, increases the critical draw ratio and therefore has a purely stabilizing effect. The critical draw ratio starts to increase moderately at $Re/Fr = \mathcal{O}(10)$ and then grows almost linearly above $Re/Fr = \mathcal{O}\left(10^2\right)$. The qualitative trend of this curve has already been shown by Shah and Pearson (1972b) and was corrected quantitatively by Chang et al. (1981), but in both cases only up to $Re/Fr = 400$, while Cao et al. (2005) and German et al. (2006) have analyzed the stability up to $Re/Fr \approx 25$. Here, a larger range was chosen as it will be shown in Chap. 9 that $Re/Fr > 400$ is generally possible in practical applications. The data presented by Chang et al. (1981) are shown in Fig. 4.1a and are in good agreement with the neutral stability curve.

It is possible to fit the critical draw ratio with an empirical expression,

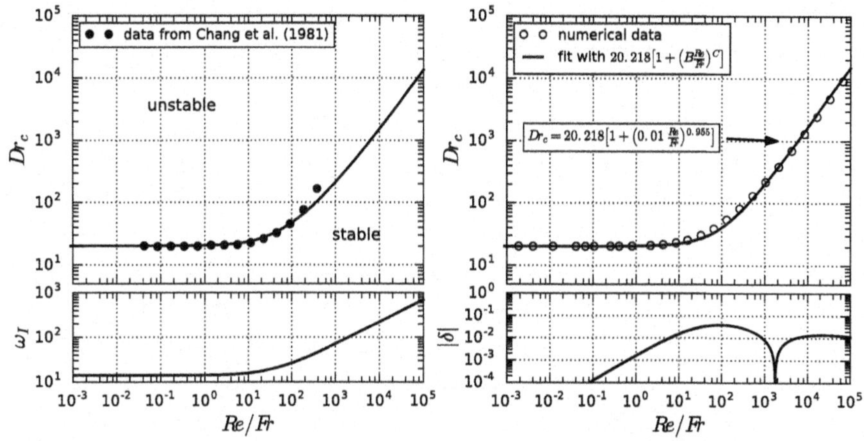

(a) Neutral stability curves.

(b) Empirical fit.

Fig. 4.1 Visualization of the stabilizing influence of gravity. **a** Neutral stability curves as calculated with numerical continuation (*lines*) and the results of Chang et al. (1981) (*points*). The parameter space below the critical draw ratio corresponds to stable processing conditions, and the space above to unstable ones. **b** Empirical fit to this data. Only few points of the numerical data are shown here for the sake of readability. The fit residual $|\delta|$ is always below 4% and the errors on the fitting parameters B and C are negligible

$$Dr_c = 20.218\left[1 + \left(B\frac{Re}{Fr}\right)^C\right], \tag{4.4}$$

with free parameters B, determined as 0.01, and C, determined as 0.955, as visualized in Fig. 4.1b. Equation (4.4) is chosen such that the numerically known critical draw ratio of the purely viscous model, $Dr_c = 20.218$, is reached when Re/Fr approaches zero. The adjusted correlation coefficient $R^2_{\text{adj}} = 1.0$ and the maximum relative deviation $|\delta|$ of 4% indicate the high accuracy of the fit. Examples of the steady states and perturbations at criticality for $Re/Fr = 10^2$ and 10^4 are shown in Fig. 4.2.

Figure 4.3 shows the perturbation of the force $F = \frac{f}{f_s} - 1$ at criticality along the x-axis for two values of Re/Fr. The impact of gravity leads to a damping effect of the oscillation amplitude close to the inlet. On the basis of the draw resonance mechanism explained in Sect. 3.3, this can be directly correlated to the increase of the critical draw ratio: An initial perturbation of the force at the outlet has a reduced impact at the inlet with increasing effect of gravity, which hinders the feedback mechanism of draw resonance. As can be seen in Fig. 4.3c, the ratio of the oscillation amplitude at the inlet to the outlet decreases nonlinearly with Re/Fr, while the stability curve in Fig. 4.1a increases almost linearly ($C \approx 1$) with Re/Fr for large values of Re/Fr. This indicates that the effect of the damped force oscillation cannot be correlated in a simple way to the critical draw ratio. Moreover, it is possible that further effects influence the propagation of the perturbation waves in a manner which cannot be seen in the ratio of the amplitudes at inlet to outlet. One possibility for such an effect

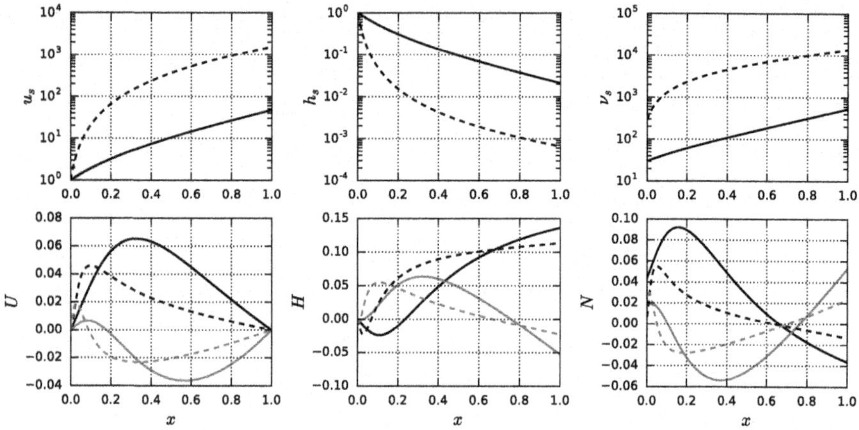

Fig. 4.2 Steady states and perturbations at criticality for the viscous–gravity film casting model. The solid lines correspond to $Re/Fr = 10^2$ and the dashed lines to $Re/Fr = 10^4$. The real parts of the complex perturbations are plotted in black and the imaginary parts in gray

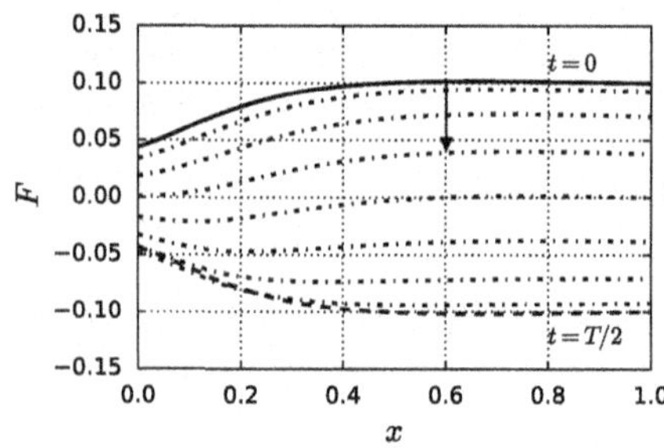

(a) $Re/Fr = 10^2$.

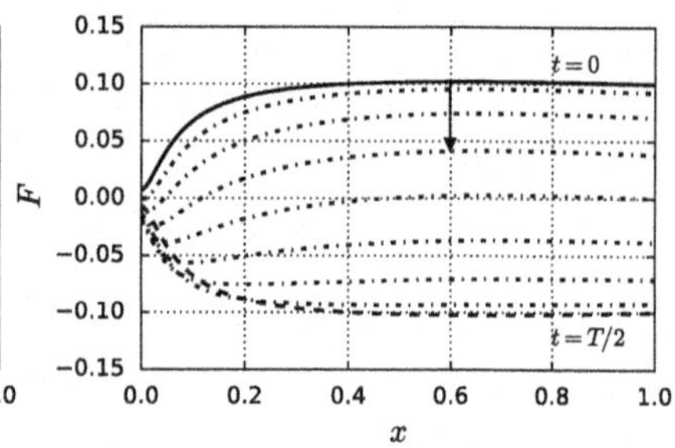

(b) $Re/Fr = 10^4$.

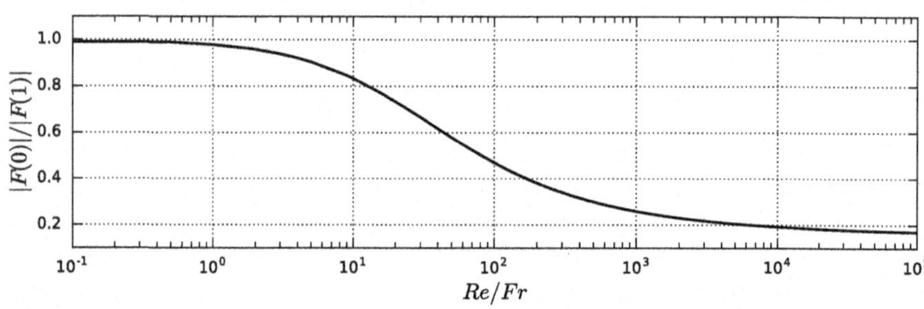

(c) Force perturbation amplitude ratio.

Fig. 4.3 **a–b** Perturbation of the force $F = f/f_s - 1$ at criticality along the x-axis for the viscous–gravity model at various times within a half period for $Re/Fr = 10^2$ and 10^4. The thick line corresponds to $t = 0$, the dashed line to $t = T/2$ and the dashed-dotted lines in between to increasing time along the arrow, with incremental step size of $T/16$. **c** Evolution of the ratio of the force perturbation amplitudes at inlet to outlet for increasing Re/Fr

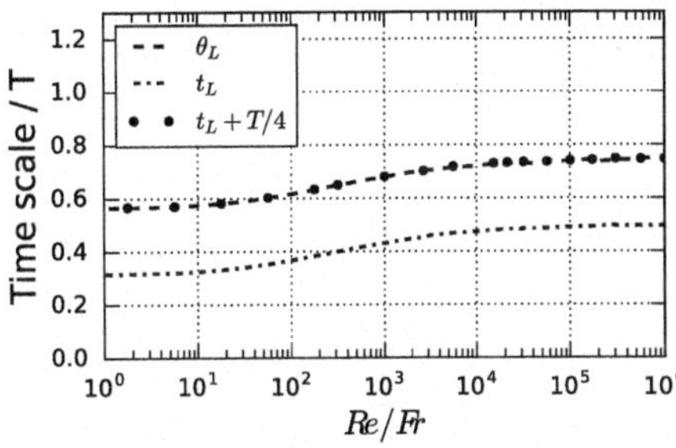

(a) Kim's criterion.

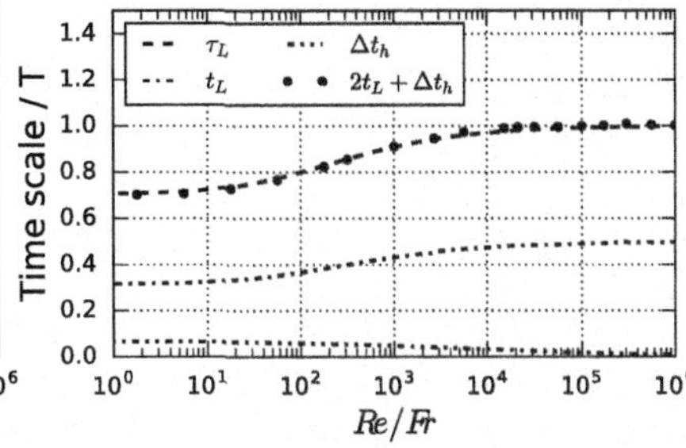

(b) Jung's criterion.

Fig. 4.4 Check of Kim's and Jung's criteria for the viscous–gravity model at criticality. The time is scaled by the oscillation period T. Only few markers are shown for $t_L + T/4$ and $2t_L + \Delta t_h$ for the sake of readability. Reproduced with permission from Bechert et al. (2016) and AIP Publishing

could be the variation of the phase of the perturbation oscillation along x, as visible in Fig. 4.3a, b, where the varying phase leads to a crossover of the constant-time lines.

In general, the stabilizing effect of gravity may not be surprising if the perturbation of the force is taken as the basis of the instability mechanism, as gravity itself constitutes a constant force along the x-axis, which is unaffected by perturbations within the fluid and therefore reduces the relative amount of the force perturbations. In order to check the validity of Kim's stability criterion (3.33) and Jung's approximative criterion (3.36), Fig. 4.4 depicts the corresponding quantities of the inequalities, calculated at criticality. It can be seen that the equality holds in both cases on the entire domain of Re/Fr, proving the validity of both criteria for the viscous–gravity model.

4.2 Influence of Inertia on Draw Resonance

The model equations of the *viscous–inertia model* for infinite-width films are similar to the ones of the viscous–gravity model. While (4.1a) and (4.1b) remain unchanged, the gravity term in (4.1c) is replaced by the inertial term (see (2.35)), yielding

$$\partial_x (h\nu) - Re\, h\, (\partial_t u + u \partial_x u) = 0 \ . \tag{4.5}$$

This changes the steady state equation (4.2c) to

$$\nu_s' = \frac{\nu_s}{4} \left(\frac{\nu_s}{u_s} + Re\, u_s \right) , \tag{4.6}$$

and the perturbation equation (4.3c) to

$$N' = -H' + Re\frac{u_s}{4}\left(1 + \frac{4\omega}{\nu_s}\right)U\,. \tag{4.7}$$

Figure 4.5a shows the dependencies of the critical draw ratio and the oscillation frequency at criticality on the Reynolds number (see Appendix B). Similar to gravity, inertia has a purely stabilizing effect. The critical draw ratio starts to increase slowly at $Re = \mathcal{O}\left(10^{-2}\right)$. Above $Re = 0.1$, the stabilizing effect increases strongly and the curve diverges at $Re_{\mathrm{us}} = 0.414$. As a consequence, the system becomes unconditionally stable if $Re > Re_{\mathrm{us}}$, i.e., no draw resonance occurs, independent of the value of Dr. The presence of this region of unconditional stability was additionally checked by an eigenvalue analysis (see Appendix A). Also shown in Fig. 4.5a are data points extracted from the stability curve presented by Hagen and Langwallner (2006), which coincide perfectly with the calculated curve here. While several authors (Shah and Pearson 1972b; Chang et al. 1981; Hagen and Langwallner 2006) have presented stability curves similar to the one shown in Fig. 4.5a, the region of unconditional stability was identified for the first time in Bechert et al. (2015).

As shown in Fig. 4.5b, the stability curve can be fitted with the following empirical correlation,

$$Dr_{\mathrm{c}} = 20.218\left[1 + \left(\frac{B\,Re}{Re_{\mathrm{us}} - Re}\right)^{C}\right], \tag{4.8}$$

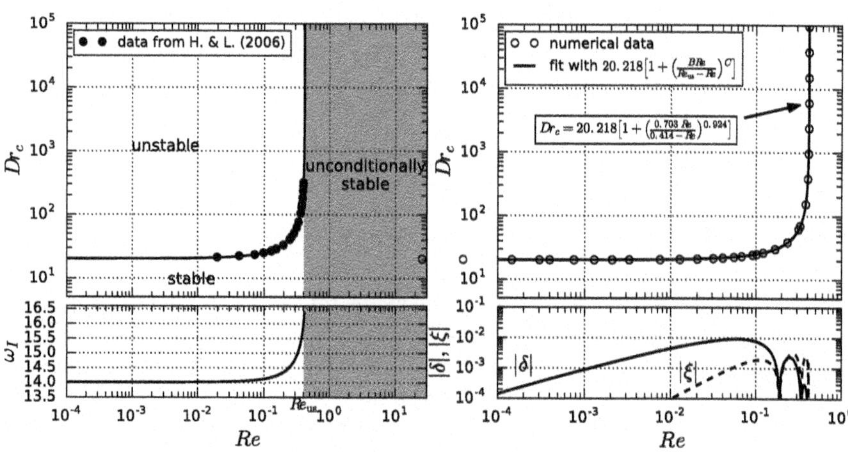

(a) Neutral stability curves.

(b) Empirical fit.

Fig. 4.5 Visualization of the stabilizing influence of inertia. **a** Neutral stability curves as calculated with numerical continuation (*lines*) and the results of Hagen and Langwallner (2006) (*points*). The parameter space below the critical draw ratio corresponds to stable processing conditions, and the space above to unstable ones. For $Re > Re_{\mathrm{us}}$, draw resonance is completely suppressed. **b** Empirical fit to the data using Deming regression. Only few points of the numerical data are shown here for the sake of readability. The fit error $|\xi|$ and the fit residual $|\delta|$ are always below 1%. The errors on the fitting parameters B, Re_{us} and C are negligible with free parameters B, determined as 0.703, Re_{us}, determined as 0.414 and C, determined as 0.924. Due to the diverging character of the stability curve, the fitting is done by Deming regression, a method which considers deviations in both variables, Re and Dr_{c} (Bechert et al. 2015). Besides the determination of the threshold point Re_{us}, (4.8) provides a practical tool for the determination of the critical draw ratio in dependence of the Reynolds number without performing any stability analysis. The accuracy of this correlation is high, the fit residual, $|\delta|$ and the fit error $|\xi|$, being below 1% for the entire range of Re. Examples of the steady states and perturbations at criticality for $Re = 0.1$ and 0.4 are shown in Fig. 4.6.

The phenomenon of unconditional stability can be understood on the basis of the steady state behavior. The steady state expressions can be derived as follows, using the flow rate conservation $h_s u_s = 1$ (see (3.6a)). This makes it possible to combine (4.1b) and (4.5) to

$$Re u_s - 4\frac{u_s'}{u_s} = -c\,, \tag{4.9}$$

with $c = c(Dr, Re)$ being a constant which needs to be determined numerically such that $u_s(1) = Dr$ is fulfilled. With the help of partial fraction decomposition, (4.9) can be integrated to

$$u_s(x) = \frac{c}{(c + Re)\,\mathrm{e}^{-cx/4} - Re}\,. \tag{4.10}$$

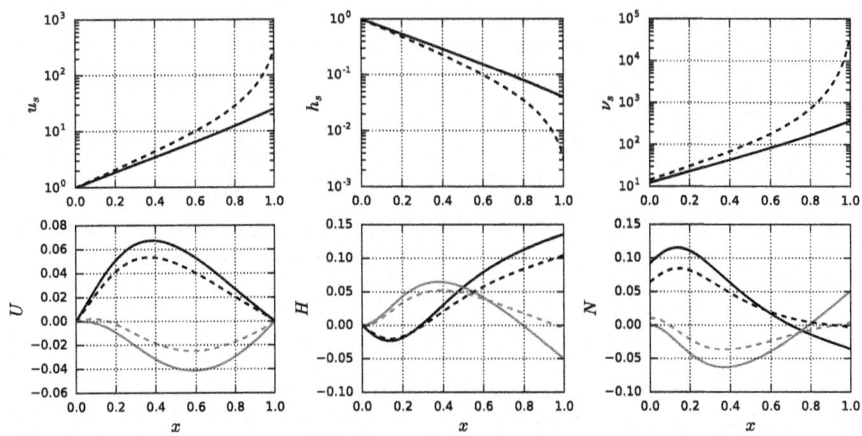

Fig. 4.6 Steady states and perturbations at criticality for the viscous–inertia film casting model. The solid lines correspond to $Re = 0.1$ and the dashed lines to $Re = 0.4$. The real parts of the complex perturbations are plotted in black and the imaginary parts in gray

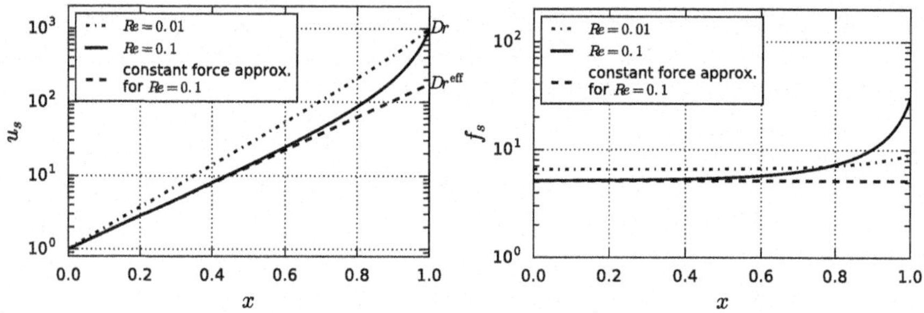

(a) Steady velocity profiles. (b) Steady force profiles.

Fig. 4.7 Steady state profiles of **a** the velocity and **b** the force for $Dr = 10^3$ and $Re = 0.01, 0.1$. The constant force approximation which leads to the effective draw ratio Dr^{eff} is shown for $Re = 0.1$. Reproduced with permission from Bechert et al. (2016) and AIP Publishing

The steady force profile can then be calculated to

$$f_s(x) = h_s u_s' = \frac{c(c + Re)}{4\left(c + Re\left(1 - \mathrm{e}^{\,cx/4}\right)\right)}\,. \tag{4.11}$$

Figure 4.7 depicts the steady velocity and force profiles for a critical draw ratio of 10^3 and two Reynolds numbers, 0.01 and 0.1. It can be seen that with increasing inertia effect, the velocity becomes smaller for the entire range of x and shows a sharp increase close to the outlet to fulfill the boundary condition. This can be related to the force profile, which is, in contrast to the purely viscous case, not constant for arbitrary values of Re, but increases along x. As the increase is located close to the outlet, the force profile remains more or less constant over a large part of the x-domain and can be roughly approximated by

$$f_s(x) \approx f_s(0) = \frac{c + Re}{4}\,. \tag{4.12}$$

This *constant force approximation* is visualized in Fig. 4.7b for $Re = 0.1$. As a consequence, the velocity profile becomes in this approximation, plugging (4.12) in (3.6b),

$$u_s = \mathrm{e}^{\,\frac{c+Re}{4}x}\,, \tag{4.13}$$

and the profile corresponding to $Re = 0.1$ is shown in Fig. 4.7a. Using (4.13), it is possible to derive an *effective draw ratio* $Dr^{\text{eff}} = Dr^{\text{eff}}(Dr, Re) = \mathrm{e}^{\,(c+Re)/4}$, which corresponds to the draw ratio as seen by the fluid along the major part of the x-axis. An approximated critical draw ratio is now defined as the draw ratio for which the corresponding effective draw ratio is equal to 20.218, i.e., the critical draw ratio of the viscous model, which has a constant force along the x-axis. This approximative

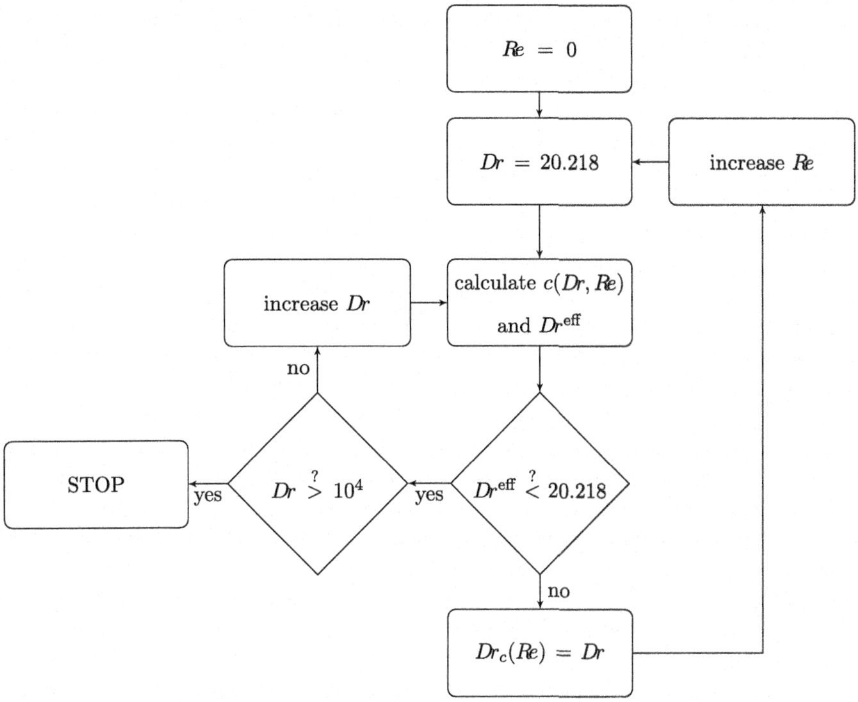

Fig. 4.8 Flow chart of the algorithm used to determine the approximative stability curve of the viscous–inertia model shown in Fig. 4.9. Reproduced with permission from Bechert et al. (2016) and AIP Publishing

critical draw ratio is calculated by stepwisely increasing the draw ratio and simultaneously calculating the effective draw ratio, until it exceeds the value of 20.218. This procedure is repeated for several Reynolds numbers and stops if the approximated critical draw ratio is above 10^4. As can be seen in Fig. 4.7a, the effective draw ratio is always equal or below the real draw ratio in the entire x-domain, therefore it is safe to start with $Dr = 20.218$ for all Reynolds numbers. The algorithm is sketched in Fig. 4.8.

The approximated stability curve is plotted in Fig. 4.9. In comparison with the curve resulting from linear stability analysis, the stability is underestimated, i.e., a higher value of Re is needed to increase the critical draw ratio. Nevertheless, this discrepancy is rather small and the qualitative behavior is perfectly recovered. This shows that the reason for the diverging stability curve and thus the unconditional stability region lies within the damping effect of inertia, which prevents an equal distribution of the acceleration along the x-axis. As a consequence, the fluid dynamics are governed by a smaller, effective draw ratio along the major part of the x-domain. If inertia is strong enough, this effective draw ratio is always below the critical value of a purely viscous (i.e., constant force) model.

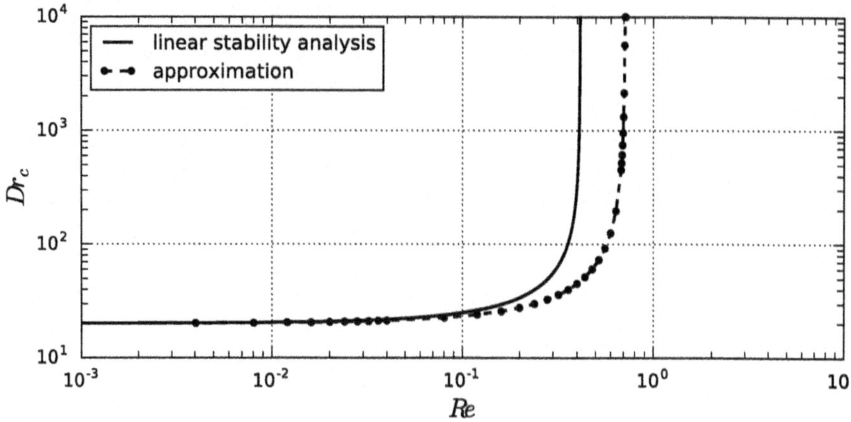

Fig. 4.9 Approximative stability curve for the viscous–inertia model obtained with the algorithm depicted in Fig. 4.8 (*dotted dashed line*) and comparison with the exact numerical solution (*solid line*). Besides an underestimation of the stabilizing influence of inertia, the essential diverging behavior is identical. Reproduced with permission from Bechert et al. (2016) and AIP Publishing

As noted above, the approximation underestimates the stabilizing effect of inertia. The effective draw ratio, however, can also rather be seen as a lower bound, as the value at $x = 0$ is taken for the constant force approximation. Consequently, the mechanism explained above using the steady state profiles can be seen as the main cause for the unconditional stability behavior, but it is not the only stabilizing mechanism induced by inertia. Similar to gravity, inertia reduces the amplitude of the force oscillations towards the inlet, as shown in Fig. 4.10, and this hinders the feedback mechanism causing draw resonance. Figure 4.10c shows that the ratio of force oscillation amplitudes at inlet to outlet decreases with Re, but does not diverge at Re_{us}. Therefore the damping of the oscillation is considered as stabilizing, but not leading to unconditional stability. Note that even close to Re_{us}, the phase of the force oscillations does not vary much along x.

Figure 4.11 shows the various traveling times and phases utilized by Kim's and Jung's stability criteria. While Fig. 4.11a displays perfect coincidence between the left-hand side and the right-hand side of (3.33) at criticality, indicating the validity of Kim's criterion in the presence of inertia for all relevant Reynolds numbers, Fig. 4.11b shows an increasing mismatch with increasing Re. Consequently, Jung's criterion (3.36) needs to be modified by an additional phase δ_{Re}, i.e.,

$$2t_L + \Delta t_h + \delta_{Re} \leq \tau_L \ , \tag{4.14}$$

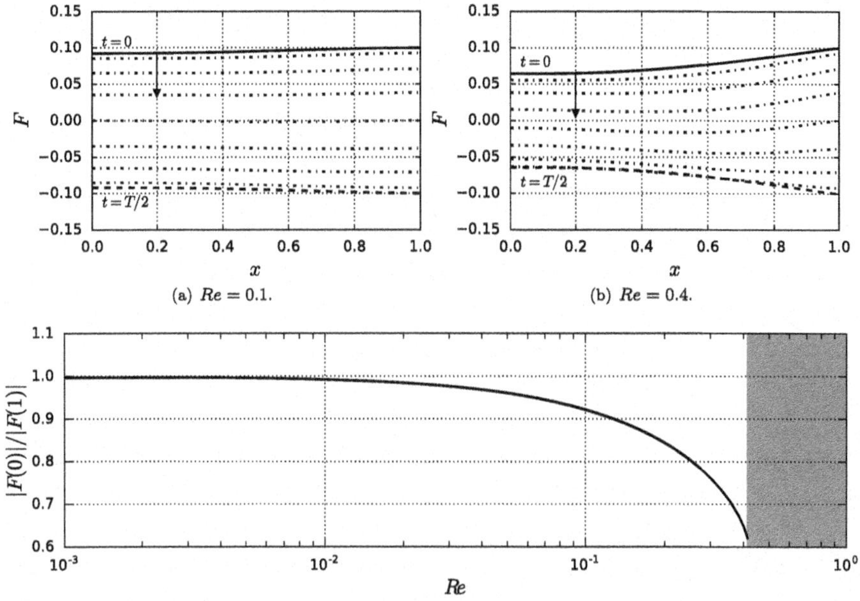

(a) $Re = 0.1$. (b) $Re = 0.4$.

(c) Force perturbation amplitude ratio.

Fig. 4.10 a–b Perturbation of the force $F = f/f_s - 1$ along the x-axis for the viscous–inertia model at various times within a half period for $Re = 0.1$ and 0.4. The thick line corresponds to $t = 0$, the dashed line to $t = T/2$ and the dashed-dotted lines in between to increasing time along the arrow, with incremental step size of $T/16$. **c** Evolution of the ratio of the force perturbation amplitudes at inlet to outlet for increasing Re. The gray region indicates unconditional stability

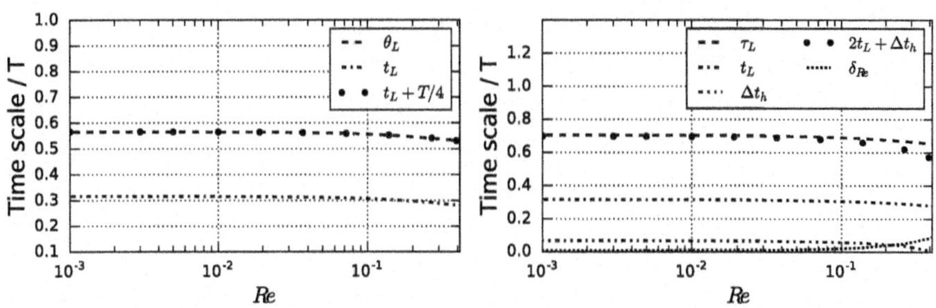

(a) Kim's criterion. (b) Jung's criterion.

Fig. 4.11 Check of Kim's and Jung's criteria for the viscous–inertia model at criticality. The time is scaled by the oscillation period T. Only few markers are shown for $t_L + T/4$ and $2t_L + \Delta t_h$ for the sake of readability. Reproduced with permission from Bechert et al. (2016) and AIP Publishing

which depends on Re and cannot be derived straightforwardly, as Jung's criterion is not based on a fundamental physical mechanism. Therefore it can be stated that in the presence of inertia, Jung's criterion is not applicable anymore.

Chapter 5
Role of Surface Tension in Fiber Drawing

Abstract In this chapter, we study the effect of surface tension on the draw resonance behavior of Newtonian fiber drawing. A region of unconditional instability caused by the influence of surface tension is found. The mechanism underlying the destabilizing effect of surface tension is discussed and established stability criteria are tested for validity in the presence of surface tension.

5.1 Influence of Surface Tension on Draw Resonance

For studying the influence of surface tension on the critical draw ratio, the *viscous–surface-tension model* for fibers is used, which corresponds to (2.16), (2.18) and (2.24) in which Re and Re/Fr are set to zero. Following the same procedure as in Sect. 3.2, using the following ansatz for the fiber area $a(x,t) = a_s(x)(1 + A(x)\exp^{\omega t})$ with $|A| \ll 1$, the steady state equations can be written as

$$a_s' = -\frac{a_s v_s}{3u_s}\,, \tag{5.1a}$$

$$v_s' = \frac{v_s}{3u_s}\left(v_s + \frac{1}{2Ca^*\sqrt{a_s}}\right)\,, \tag{5.1b}$$

$$u_s' = \frac{v_s}{3}\,, \tag{5.1c}$$

and the perturbation equations as

$$A' = -\left(\frac{\omega}{u_s}A + U'\right)\,, \tag{5.2a}$$

$$N' = -A' - \frac{1}{2Ca^*\sqrt{a_s}}\left[\frac{1}{6u_s}(A + 2N) + \frac{A'}{v_s}\right]\,, \tag{5.2b}$$

$$U' = \frac{v_s}{3u_s}(N - U)\,. \tag{5.2c}$$

© Springer Nature Switzerland AG 2025

M. Bechert and B. Scheid, *Draw Resonance*, Fluid Mechanics and Its Applications 141,
https://doi.org/10.1007/978-3-031-85568-9_5

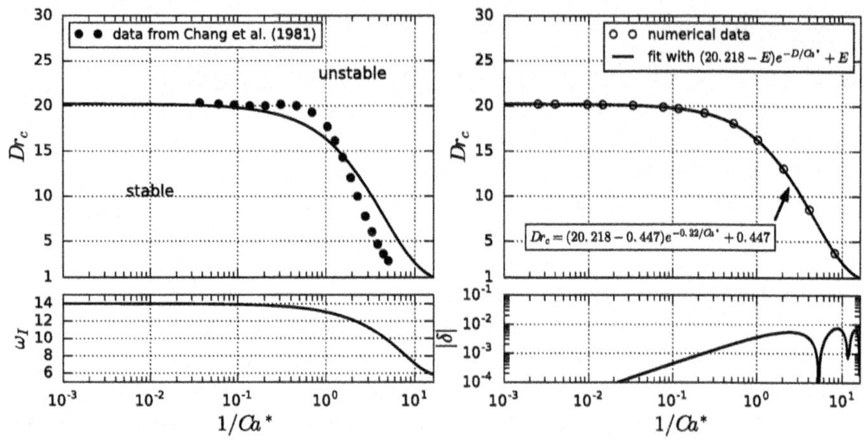

(a) Neutral stability curves.

(b) Empirical fit.

Fig. 5.1 Visualization of the destabilizing influence of surface tension. **a** Neutral stability curves as calculated with numerical continuation (*lines*) and the results of Chang et al. (1981) (*points*). The parameter space below the critical draw ratio corresponds to stable processing conditions, and the space above to unstable ones. **b** Empirical fit to this data. Only few points are shown here for the sake of readability

The neutral stability behavior is visualized in Fig. 5.1a. Surface tension clearly has a purely destabilizing effect on draw resonance. In particular, for $1/Ca^* > 10$ the destabilization is so strong that the critical draw ratio reaches unity, which means that no stable acceleration of the fiber is possible at all. Similar curves have been presented by Shah and Pearson (1972b) and Chang et al. (1981), the latter also depicted in the plot. The moderate agreement between these results and the present curve, as well as the apparent non-monotonous stability behavior, are most likely caused by the rather low quality of the original plot as presented by Chang et al. (1981). As depicted in Fig. 5.1b it is possible to fit the critical draw ratio with high accuracy using the following empirical expression,

$$Dr_c = (20.218 - E)\,e^{-\frac{D}{Ca^*}} + E\ , \tag{5.3}$$

with free parameters D, determined as 0.220 ± 0.000, and E, determined as 0.447 ± 0.002. The adjusted correlation coefficient $R^2_{\text{adj}} = 1.0$ and the maximum relative deviation $|\delta|$ of 4% indicate the high accuracy of the fit. Using this correlation, a critical capillary number for which $Dr_c = 1$ can be calculated as $1/Ca^*_c = 16.26$. Above this value, the system becomes *unconditionally unstable*, as no draw ratio equal or above unity exists, for which stable drawing is possible. The presence of this region was verified by eigenvalue spectrum method (Appendix A). Examples of the steady states and perturbations at criticality for $1/Ca^* = 5$ and 10 are shown in Fig. 5.2.

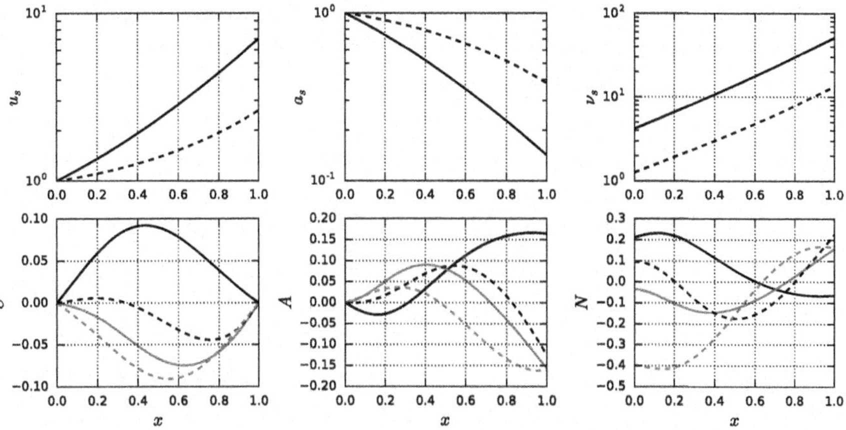

Fig. 5.2 Steady states and perturbations at criticality for the viscous–surface-tension fiber drawing model. The solid lines correspond to $1/Ca^* = 5$ and the dashed lines to $1/Ca^* = 10$. The real parts of the complex perturbations are plotted in black and the imaginary parts in gray

5.2 Destabilization Mechanism of Surface Tension

In order to investigate the destabilizing effect further, Fig. 5.3a,b show the time evolution of the force perturbations for $1/Ca^* = 5$ and 10. In contrast to the previously discussed effects of gravity and inertia, surface tension leads to an increase of the oscillation amplitude towards the inlet. Figure 5.3c visualizes the ratio of the amplitudes at inlet to outlet. While the ratio increases with increasing $1/Ca^*$ for $1/Ca^* < 9$, it reaches a maximum in the vicinity of the inflection point of the neutral stability curves shown in Fig. 5.1a and decreases for larger values, which seems to contradict the interpretation of the feedback mechanism. However, as can be seen in Fig. 5.3b, the phase of the force oscillations varies strongly along x, which most likely has an additional influence on the propagation of the perturbation waves along x.

The destabilizing influence of surface tension can also be investigated on the basis of the *Plateau–Rayleigh instability* mechanism.[1] The most unstable mode of the Plateau–Rayleigh instability is approximately given by 9.01 R (Charru, 2011),[2] which is definitely smaller than the drawing length L. In general, the radially acting

[1] The Plateau–Rayleigh instability is responsible for the breaking of a liquid thread into a train of droplets.

[2] In the Plateau-Rayleigh instability, what sets the critical wavelength below which the liquid thread is stable is the longitudinal curvature that has a stabilizing effect, as opposed to the radial curvature that plays a destabilizing role. Note that the longitudinal curvature has been neglected in the fiber model as it appeared in the higher-order of the asymptotical expansion based on the smallness of the aspect ratio $\alpha = R/L \ll 1$ (see Sect. 2.1).

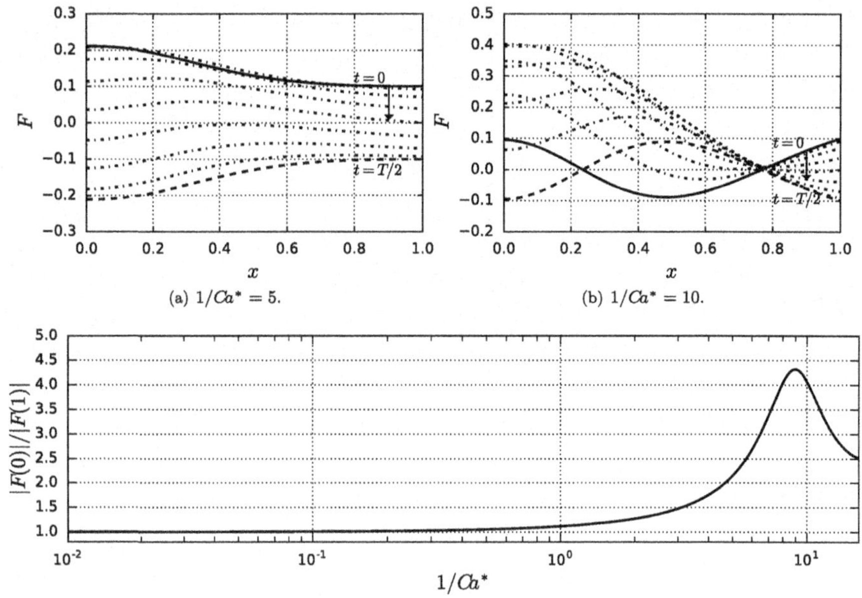

(a) $1/Ca^* = 5$.

(b) $1/Ca^* = 10$.

(c) Force perturbation amplitude rate.

Fig. 5.3 a–b Perturbation of the force $F = f/f_s - 1$ along the x-axis for the viscous–surface-tension fiber model at various time values within a half period for $1/Ca^* = 5$ and 10. The thick line corresponds to $t = 0$, the dashed line to $t = T/2$ and the dashed-dotted lines in between to increasing time along the arrow, with incremental step size of $T/16$. **c** Evolution of the ratio of the force perturbation amplitudes at inlet to outlet for increasing $1/Ca^*$

surface tension force enforces perturbations in the fiber. This effect is more pronounced at the inlet, where the capillary pressure is minimal (see Eq. (2.19)), as compared to the outlet, where the capillary pressure is maximal due to the smaller fiber radius. This mechanism is therefore consistent with the force perturbation amplitude ratio shown in Fig. 5.3c.

The alternative stability criteria of Kim and Jung are checked with the plots of Fig. 5.4. Instead of maximum- and minimum-thickness waves, the corresponding cross-section area waves are considered and Δt_a is used instead of Δt_h. Similar to the previously analyzed effects, Kim's criterion is also valid in the presence of surface tension effects, as Fig. 5.4a shows perfect coincidence of both sides of (3.33) at criticality. Figure 5.4b, however, reveals a mismatch of both sides of Jung's criterion (3.36) and an additional phase δ_{Ca^*}, also depicted in Fig. 5.4b, needs to be added, analogue to the modified criterion (4.14) of the viscous–inertia model. It can be seen, that the phase difference Δt_a reaches almost $T/2$ for large values of $1/Ca^*$, i.e., the cross-section perturbations and the force perturbations are almost in phase

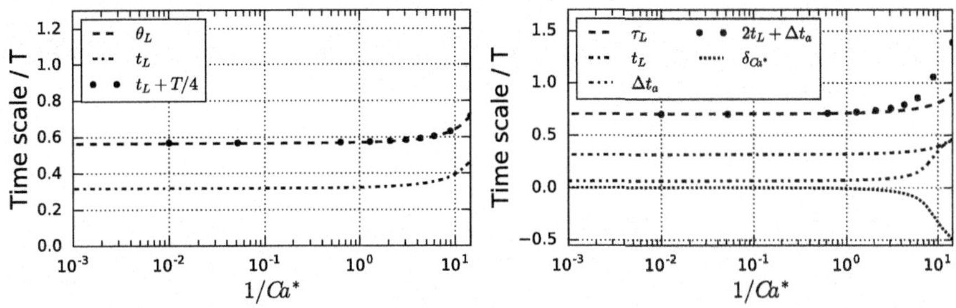

(a) Kim's criterion. (b) Jung's criterion.

Fig. 5.4 Check of Kim's and Jung's criteria for the viscous–surface-tension fiber model at criticality. The time is scaled by the oscillation period T. Only few markers are shown for $t_L + T/4$ and $2t_L + \Delta t_a$ for the sake of readability

opposition. The phase variation shown in Fig. 5.3b also indicates a strong dependence of the phase on x. It is supposed that the approximation used to obtain Jung's criterion is not valid anymore in the case of such a strong phase variation.

Chapter 6
Finite Width Effects in Film Casting

Abstract In this chapter, we investigate the influence of the reduction of width along the stretching direction, the so-called neck-in effect, on the draw resonance instability in Newtonian film casting problem using a linear stability analysis of a model of reduced dimensionality. We introduce the local Trouton ratio as a measure for the type of elongational deformation, which can be uniaxial, planar, or a combination of both. In the case of purely uniaxial or planar deformations, a one-dimensional model is sufficient. Additionally, the stabilizing mechanisms of the neck-in is revealed. Known alternative stability criteria are extended to the case of finite width and their validity is tested for finite aspect ratios.

6.1 Geometrical Effects

If the assumption of infinite width cannot be applied, e.g., if the casting length is of similar order as the initial film width, further geometrical effects come into play and the analogy to fiber drawing is not valid anymore. A two- or even three-dimensional modeling of the film casting process is already challenging if only steady state solutions are sought and numerous works exist on this topic (e.g., d'Halewyu et al. 1990; Sakaki et al. 1996; Smith and Stolle 2000a). However, as the focus of this book lies primarily on the stability of the process, mainly works directly linked to draw resonance will be discussed in the following. Besides draw resonance, two phenomena[1] can generally be observed in film casting. While the *neck-in* effect leads to a reduction of the film width along the casting axis, the *edge-bead* or *dog-bone* effect results in a larger film thickness at the edges compared to the center. In fact, both effects are primarily[2] caused by a combination of two different deformation types (Dobroth and

[1] These phenomena are also often called instabilities, but this term will not be used here to avoid confusion. Technically, they could be seen as *steady* instabilities, to distinguish them from the *transient* instability of draw resonance.

[2] Surface tension may additionally lead to edge-bead (Dobroth and Erwin 1986).

© Springer Nature Switzerland AG 2025

M. Bechert and B. Scheid, *Draw Resonance*, Fluid Mechanics and Its Applications 141,
https://doi.org/10.1007/978-3-031-85568-9_6

Erwin 1986; Ito et al. 2003). While the center part of the film is dominated by *planar* elongational deformation, *uniaxial* elongational deformation dominates at the edges. Planar elongation is characterized by purely parallel streamlines. In uniaxial elongation, the stress-free edges lead to a reduction of width.

Silagy et al. (1996) studied the influence of neck-in on draw resonance for the first time. They extended the one-dimensional model of Yeow (1974) such that the film width can vary along the casting axis. By additionally proposing a linear transversal velocity profile, they could reduce the two-dimensional model to a one-dimensional model with finite width. A linear stability analysis revealed a stabilizing influence of neck-in and an optimal value at which the critical draw ratio is maximal. In a succeeding paper (Silagy et al. 1998), they analyzed the stability of a full two-dimensional model without a priori defined transversal velocity profile, which made it possible to account for the edge-bead effect as well. The qualitative relationship between the critical draw ratio and the aspect ratio remained the same, but the location of the stability maximum was shifted and the stabilizing effect appeared to be increased by the edge-bead effect as well. Due to the higher complexity of the two-dimensional model, the stability was only investigated in terms of full transient solutions instead of a linear stability analysis. Later, Scheid et al. (2009b) analyzed the change in stability if either the lateral stress or the velocity at the film edges is prescribed by adapting the reduced one-dimensional model of Silagy et al. (1996).

Ahmed and Khayat (2012) performed a three-dimensional linear stability analysis of the film casting process. They reduced the steady state to the one-dimensional limit of infinite width, though, neglecting neck-in and edge-bead effects and letting only the time-dependent variations remain three-dimensional. Additionally accounting for gravity and inertia effects, they observed a stability enhancement caused by the three-dimensional perturbations. The stability behavior under the influence of neck-in, gravity and inertia was investigated extensively, which will be presented in Chap. 9.

The membrane approximation may become inappropriate if the gap between die and take-up is chosen to be small, close to the order of the initial film thickness. Kallel et al. (2015) have analyzed this particular case using different numerical approaches and detected a strong stabilizing effect with decreasing ratio of casting length to initial thickness.

6.2 Finite Width Film Model

While the one-dimensional film casting model presented in Sect. 2.2 can be applied to several experimental setups, it fails to predict the reduction of film width, the so-called neck-in, which appears if the film width is not large as compared to the casting length. In order to circumvent this drawback, a reduced one-dimensional model, which accounts for the finite width of the film, will be derived within this section by following the ansatz presented by Silagy et al. (1996), while keeping inertia and gravity effects.

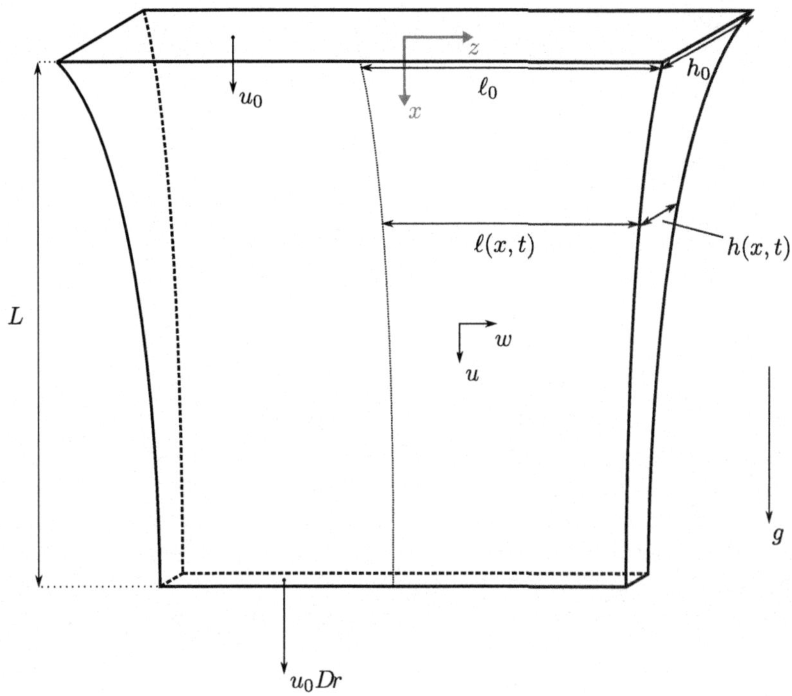

Fig. 6.1 Sketch of the finite-width film casting model. Adapted with permission from Bechert et al. (2016) and AIP Publishing

Figure 6.1 shows the sketch of the film casting model including neck-in. The film width is given by 2ℓ, and the aspect ratio $A = \ell_0/L$ is introduced as an additional control parameter,[3] with initial half film width ℓ_0. The transversal velocity component is denoted by w and the scaling rules (2.28) are complemented by $z \to \ell_0\, z$, $\ell \to \ell_0\, \ell$ and $w \to u_0\, w$. Analogue to the one-dimensional case shown in Sect. 2.2, the depth-averaged governing equations can be derived by integrating the continuity equation (2.1) and the momentum equation (2.2) over the thickness, using the extended stress boundary condition with normal vector $\mathbf{n}_\pm = (-\partial_x h_\pm \mathbf{e}_x + \mathbf{e}_y - \partial_z h_\pm \mathbf{e}_z)/\sqrt{1 + 1/4\left((\partial_x h)^2 + (\partial_z h)^2\right)}$ and being aware that $\partial_y u = 0$ is still true at leading order. The scaled continuity equation then becomes at leading order

$$\partial_t h + \partial_x(hu) + \frac{1}{A}\partial_z(hw) = 0\,, \tag{6.1}$$

[3] In some cases the aspect ratio is defined as L/ℓ_0 (e.g., Silagy et al. 1996; Kim et al. 2005; Shin et al. 2007).

and the scaled x- and z-components of the momentum equation read

$$Re\, h\,(\partial_t u + u\partial_x u + w\partial_z u) = \partial_x\,(h\sigma_{xx}) + \frac{1}{A}\partial_z(h\sigma_{xz}) + h\frac{Re}{Fr}\,, \tag{6.2a}$$

$$Re\, h(\partial_t w + u\partial_x w + w\partial_z w) = \partial_x(h\sigma_{xz}) + \frac{1}{A}\partial_z(h\sigma_{zz})\,. \tag{6.2b}$$

In order to simplify the analysis further, the equations above will be reduced to a one-dimensional model by posing the following ansatz (Narayanaswamy 1977; Silagy et al. 1996):

$$u = u(x,t),\quad h = h(x,t),\quad w = z\frac{w_\ell(x,t)}{\ell(x,t)}\,, \tag{6.3}$$

i.e., both the axial velocity and the thickness do not vary along the film width[4] and the transverse velocity profile is linear in z, with edge velocity w_ℓ. It is not possible to simplify the system in a similar way as it is done in Chap. 2, i.e., by using A as a small order parameter and expand the equations, as the entire range of aspect ratios will be analyzed here. The assumption of a linear transverse velocity profile can be justified at least for linear polypropylenes by measurements from Grieß (2014). Making the thickness constant along the width basically implies neglecting the edge-bead effect.

Integrating (6.1) along half the film width, with the help of the kinematic condition,

$$A(\partial_t\ell + u\partial_x\ell) - w_\ell = 0\,, \tag{6.4}$$

directly yields

$$\partial_t(h\ell) + \partial_x(h\ell u) = 0\,. \tag{6.5}$$

The vector normal to the boundary surface is given by $\mathbf{m} = (-A\partial_x\ell\mathbf{e}_x + \mathbf{e}_z)/\sqrt{1+(A\partial_x\ell)^2}$, and the stress boundary conditions are

$$-A\partial_x\ell\sigma_{xx}|_\ell + \sigma_{xz}|_\ell = 0\,, \tag{6.6a}$$

$$-A\partial_x\ell\sigma_{xz}|_\ell + \sigma_{zz}|_\ell = 0\,, \tag{6.6b}$$

which can be rearranged to write

$$\sigma_{zz}|_\ell = (A\partial_x\ell)^2\sigma_{xx}|_\ell\,. \tag{6.7}$$

Recalling that $\sigma_{yy} = 0$ at leading order and thus that the pressure p is equal to $2\partial_y v = -2\,(\partial_x u + w_\ell/(A\ell))$, the scaled constitutive equation (1.7) can be rewritten as

[4] As mentioned before, the axial velocity does not vary along the thickness due to the *membrane approximation*.

$$\sigma_{xx} = 4\partial_x u + 2\frac{w_\ell}{A\ell} , \tag{6.8a}$$

$$\sigma_{zz} = 2\partial_x u + 4\frac{w_\ell}{A\ell} , \tag{6.8b}$$

$$\sigma_{xz} = z\partial_x \left(\frac{w_\ell}{\ell}\right) . \tag{6.8c}$$

As both σ_{xx} and σ_{zz} do not depend on z, (6.7) is true over the entire domain. Together with (6.8a) and (6.8b), the relevant part of the constitutive equation can be reduced to

$$\sigma_{xx} = \nu = \tau_{xx} - \tau_{yy} = Tr(x,t)\partial_x u , \tag{6.9a}$$

$$\text{with } Tr(x,t) = \frac{6}{2 - (A\partial_x \ell)^2} \tag{6.9b}$$

being the *local Trouton ratio*. While the Trouton ratio is constant for the purely one-dimensional models, i.e., 3 for fiber drawing, see Sect. 2.1, and 4 for infinite-width film casting, see Sect. 2.2, it depends on the particular shape of the film width here and therefore on x. In the case of small aspect ratios, i.e., $A = \mathcal{O}(\beta)$, (6.9b) yields $Tr \to 3$ at leading order. In the limit of infinite width, i.e., $A \to \infty$ and $\partial_x \ell \to 0$, Eqs. (6.7–6.8b) lead to $(A\partial_x \ell)^2 \to 1/2$ and consequently $Tr \to 4$. For this reason, these limits are referred to as the *uniaxial limit* and the *planar limit*, respectively. The non-constant nature of the Trouton ratio in (6.9b) indicates the mixture of both uniaxial and planar elongational deformation.

Equation (6.2a) can be integrated along half the film width, using (6.6a), to get

$$Re\, h\ell(\partial_t u + u\partial_x u) = \partial_x(h\ell\nu) + \frac{Re}{Fr} h\ell . \tag{6.10}$$

To close the model, the kinematic condition (6.4) is rearranged, using (6.8a) and (6.9a), to yield

$$\partial_t \ell + u\partial_x \ell + \frac{\ell}{2}\partial_x u(4 - Tr) = 0 . \tag{6.11}$$

Finally, the *finite-width film model* equations (6.5), (6.9), (6.10) and (6.11) are closed with the following boundary conditions,

$$h(x=0,t) = u(x=0,t) = \ell(x=0,t) = 1 , \tag{6.12a}$$

$$u(x=1,t) = Dr . \tag{6.12b}$$

In the uniaxial limit ($Tr = 3$), the model reduces to the fiber model with neglecting surface tension, i.e. (2.16), (2.18) and (2.24) with $1/Ca^* = 0$, in which the cross-section area is given by $a = 2h\ell$. In the planar limit ($Tr = 4$), the model reduces to the infinite-width film model, i.e. Eqs. (2.33–2.35), as the film width remains constant in steady conditions according to (6.11) and can consequently be ruled out.

6.3 Influence of Neck-in on Draw Resonance

The finite-width film model presented in section Sect. 6.2 is used to investigate the influence of neck-in on the film casting stability. As the focus lies primarily on neck-in, both Re and Re/Fr are set to zero in (6.10), yielding the *viscous–neck-in model.* As for the other variables in (3.24), the additional variable ℓ is decomposed into a steady state ℓ_s and a complex perturbation L, as follows,

$$\ell(x,t) = \ell_s(x)\left(1 + L(x)\,\mathrm{e}^{\omega t}\right), \quad \text{with } L \ll 1\,, \tag{6.13}$$

the boundary conditions (3.11c) are complemented by

$$L(0) = 0\,, \tag{6.14}$$

and the condition (3.28) for the perturbation amplitude is modified to

$$H(1) + L(1) + N(1) = 0.1\,, \tag{6.15}$$

as the corresponding force is here given by $h\ell v$.

The steady state equations are

$$\ell_s' = \frac{3u_s - \sqrt{9u_s^2 + 2A^2\ell_s^2 v_s^2}}{2A^2\ell_s v_s}\,, \tag{6.16a}$$

$$u_s' = \frac{v_s}{6}\left(2 - A^2\ell_s'^2\right)\,, \tag{6.16b}$$

$$h_s' = -h_s\left(\frac{\ell_s'}{\ell_s} + \frac{u_s'}{u_s}\right)\,, \tag{6.16c}$$

$$v_s' = v_s\frac{u_s'}{u_s}\,. \tag{6.16d}$$

Due to the occurrence of $\ell_s'^2$ in (6.16a) and (6.16b), there exist in general two solutions for ℓ_s'. The second was discarded, though, as it is assumed that ℓ_s cannot monotonically grow along x under pulling conditions.

The perturbation equations are

$$L' = c_1 U + c_2 L + c_3 N \ , \tag{6.17a}$$

$$U' = \frac{u_s'}{u_s}(N - U) - \frac{\nu_s}{3u_s} A^2 \ell_s' \left(\ell_s' L + \ell_s L'\right) \ , \tag{6.17b}$$

$$H' = -\left[\frac{\omega}{u_s}(H + L) + U' + L'\right] \ , \tag{6.17c}$$

$$N' = -(H' + L') \ , \tag{6.17d}$$

where the following substitutions have been made for the sake of readability,

$$c_1 = \frac{3u_s \ell_s'}{\ell_s(2A^2 \nu_s \ell_s \ell_s' - 3u_s)} \ , \tag{6.18a}$$

$$c_2 = 3\frac{\ell_s\left(\omega + \frac{\nu_s}{6}\right) + \ell_s'\left(u_s - A^2 \nu_s \ell_s \ell_s'\right)}{\ell_s(2A^2 \nu_s \ell_s \ell_s' - 3u_s)} \ , \tag{6.18b}$$

$$c_3 = \frac{\nu_s(1 - 2A^2 \ell_s'^2)}{4A^2 \nu_s \ell_s \ell_s' - 6u_s} \ . \tag{6.18c}$$

The system of Eqs. (6.16) and (6.17) is solved by continuation using (3.11c), (6.12) and (6.14) as boundary conditions. Figure 6.2a depicts the influence of the aspect ratio on the critical draw ratio Dr_c and the oscillation frequency at criticality ω_I. The critical draw ratio curve is consistent with the part presented by Silagy et al. (1996) and the curve shown by Scheid et al. (2009b). A stability maximum is reached at $A = 0.66$, where $Dr_c = 35.1$. For $A \ll 1$ and $A \gg 1$, i.e., the uniaxial and planar limits (see section Sect. 6.2), Dr_c is identical to 20.218, the value corresponding to the one-dimensional viscous fiber and film models. The Trouton ratio at steady state and criticality[5] Tr_c, as shown in Fig. 6.2(b), increases monotonically from 3 to 4 with increasing aspect ratio. For $3 < Tr < 4$, both uniaxial and planar deformations superimpose, and the Trouton ratio can be used as a measure for the type of deformation. In general, the Trouton ratio also depends on x and decreases monotonically from inlet to outlet. As explained below, this is crucial for the stabilizing effect to occur. Close to the outlet, Tr_c increases steeper with increasing aspect ratio as compared to the inlet. This can be explained by the neck-in effect, as the reduction of width along x leads to a decreasing ratio of film width to casting length, which increases the amount of uniaxial deformation and consequently lowers the value of the Trouton ratio.

The amount of neck-in can be quantified by the *neck-in parameter* ζ, defined as

$$\zeta = 1 - \ell|_{x=1} \ , \tag{6.19}$$

[5] As the time-dependent perturbations of Tr appear to be very small and will not be discussed throughout this book, Tr_c always denotes the steady state value.

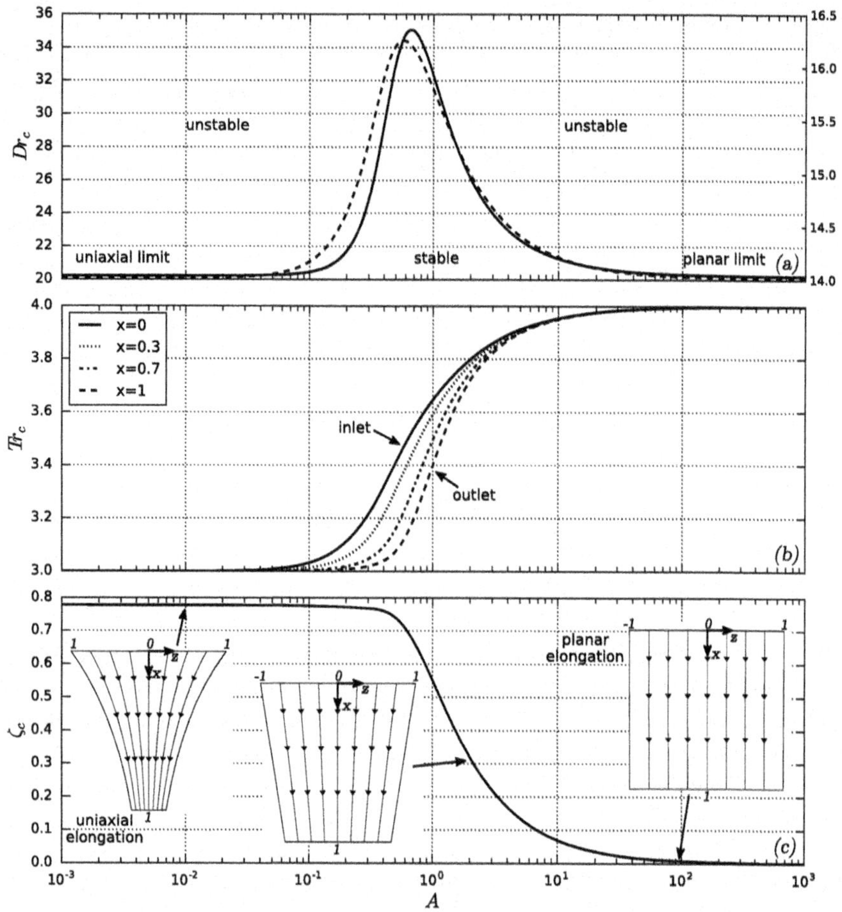

Fig. 6.2 **a** Neutral stability curves for the *viscous–neck-in model*. The solid line indicates the critical draw ratio Dr_c and the dashed line the dimensionless frequency at criticality ω_I. **b** Dependence of the steady state Trouton ratio at criticality on the aspect ratio for various positions on the x-axis. The transition from the inlet to the outlet for a fixed aspect ratio appears to be monotonic. **c** Dependence of the neck-in parameter at criticality on the aspect ratio together with typical streamline plots for $A = 0.01, 2$ and 94. Adapted with permission from Bechert et al. (2016) and AIP Publishing

which is bounded between 0 and 1. Figure 6.2c plots the neck-in parameter at criticality over the domain of aspect ratios, together with typical streamline plots at three distinct regimes of deformation. In the planar limit, no neck-in occurs and $\zeta_c = 0$. With decreasing aspect ratio, ζ_c increases, until it reaches a plateau value of $1 - Dr_c^{-1/2} = 0.7776$ in the uniaxial limit[6], where $Tr_c = 3$. As mentioned in Sect. 6.2, the neck-in model reduces to the one-dimensional fiber and film models if the

[6] The stationary expression of (6.5) gives $h_s \ell_s u_s = 1$, whereas in the uniaxial limit, $h_s = \ell_s$, such as $\ell(1) = 1/\sqrt(Dr)$.

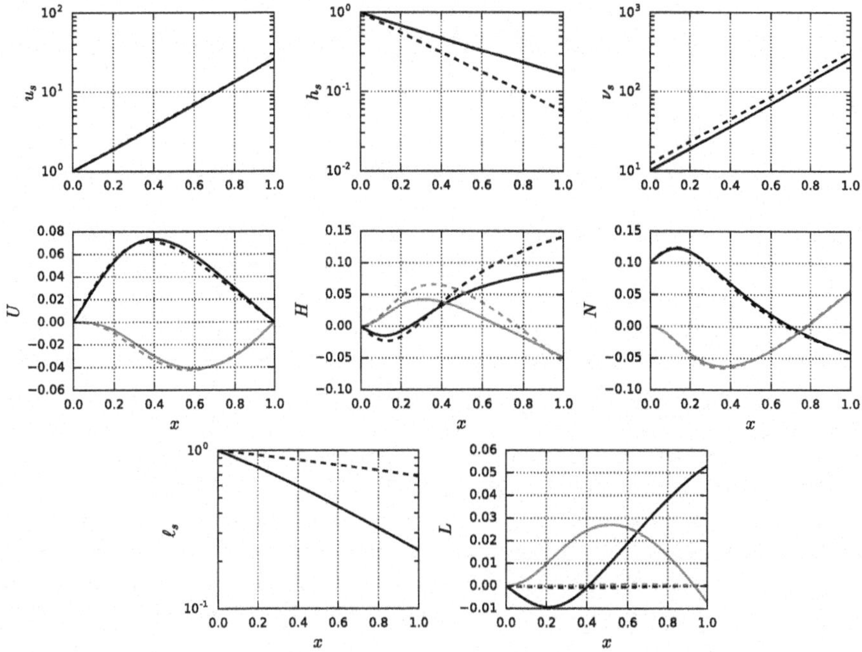

Fig. 6.3 Steady states and perturbations at $Dr_c = 26$ for the *viscous–neck-in film model*. The solid lines correspond to $A = 0.328$ and the dashed lines to $A = 2.018$. The real parts of the complex perturbations are plotted in black and the imaginary parts in gray

Trouton ratio becomes 3 or, respectively, 4. Consequently, it can be inferred from Fig. 6.2b that neck-in influences the stability only for $0.1 < A < 10$, as Tr_c deviates from the constant value by less than 1% outside this range.

Figure 6.3 depicts the steady states and perturbations at $A = 0.328$ and 2.018, where $Dr_c = 26$. The steady state velocity and the perturbations of velocity and normal stress difference coincide in both cases and the steady state normal stress difference is slighlty shifted, which can be explained by the different Trouton ratios. While the smaller aspect ratio is closer to the uniaxial regime, the larger value approaches already the planar regime. This can be seen in the similar trends of the steady states and perturbations of thickness and width for $A = 0.328$, whereas the change of ℓ_s along x and the absolute value of L are clearly less pronounced for $A = 2.018$.

In contrast to inertia, gravity and surface tension, the influence of neck-in on the critical draw ratio cannot be further analyzed and explained by examination of the force perturbations along the x-axis, as (6.10) directly leads to a force independent of x (for $Re = 0$) and thus the ratio of force perturbations at inlet to outlet is always equal to unity. Instead, (3.31) needs to be modified to account for the finite width. From continuity equation (6.5) and the momentum equation (6.10) in which $Re = 0$, the following relation can be derived using (6.9a),

$$(\partial_x (h\ell))^{\text{in}} = -\frac{Tr^{\text{out}}}{Tr^{\text{in}}} (h\ell\partial_x u)^{\text{out}} , \tag{6.20}$$

with superscripts indicating the positions at inlet and outlet. Similar to (3.31), (6.20) relates the variation of the cross-section at the inlet (left-hand side) to changes in the flow variables at the outlet (right-hand side). The ratio of Trouton ratios at outlet to inlet tunes hereby the magnitude of a perturbation at the inlet, which is induced by a perturbation at the outlet. In other words, for $Tr^{\text{out}}/Tr^{\text{in}} < 1$, the system is expected to be more stable than the one-dimensional viscous model, while it should be less stable if the ratio is larger than unity. This hypothesis is affirmed in Fig. 6.4, where the ratio of Trouton ratios has a minimum value at $A = 0.58$, which is very close to the stability maximum at $A = 0.66$. This reveals that the main reason for the stabilizing effect of neck-in lies within the difference of Trouton ratios along the x-axis, especially at inlet and outlet, as a constant force then leads to different amplitudes in the perturbation of the flow variables. As already mentioned, the time-dependent perturbation part of the Trouton ratio is negligibly small. The small mismatch between the two extrema in Fig. 6.4 could be explained by further unknown, minor stabilizing mechanisms which have most effect at slightly different aspect ratios. Nevertheless, the analysis of the steady state Trouton ratios reveals that most of the stabilizing effect accompanying neck-in is based on the steady state behavior, and barely on time-dependent effects.

The differences in the model equations of the finite-width film model as compared to the infinite-width film model necessitate an extension or modification of Kim's and Jung's stability criteria. Following the derivation presented by Kim et al. (1996), it can be deduced that the maximum or minimum thickness perturbation wave traveling time θ_L appearing in Kim's criterion (3.33) needs to be replaced by the traveling time of a maximum or minimum cross-section area wave, which belongs to the perturbation $H + L$. This time is still denoted by θ_L for the sake of simplicity. Analogue to this, the time shift Δt_h appearing in Jung's criterion (3.36) is replaced by

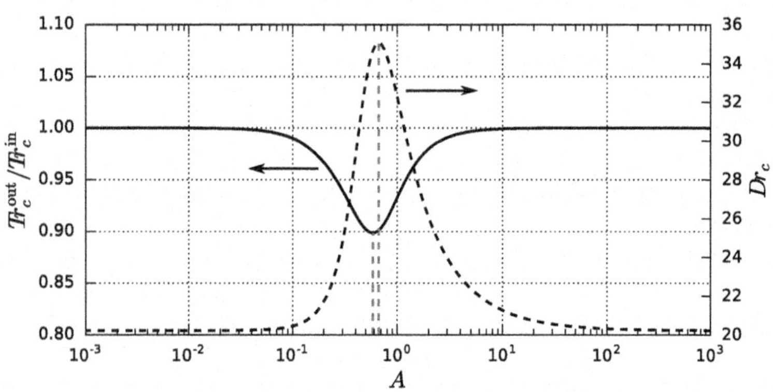

Fig. 6.4 Ratio of the critical steady state Trouton ratios at outlet to inlet over the domain of aspect ratio. Close to the stability maximum at $A = 0.66$, this ratio becomes minimal at $A = 0.58$. Reproduced with permission from Bechert et al. (2016) and AIP Publishing

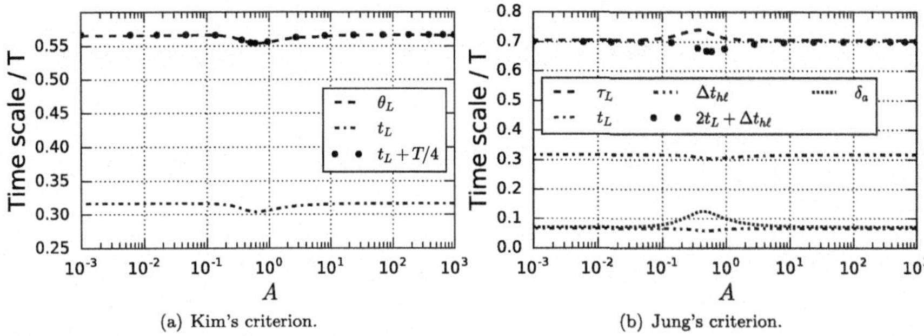

Fig. 6.5 Check of modified Kim's and Jung's criteria for the viscous–neck-in model at criticality. The time is scaled by the oscillation period T. Only few markers are shown for $t_L + T/4$ and $2t_L + \Delta t_{h\ell}$ for the sake of readability. Reproduced with permission from Bechert et al. (2016) and AIP Publishing

$\Delta t_{h\ell}$, the phase difference between the force and the cross-section area perturbations at the outlet. Figure 6.5a shows the validity of the modified version of Kim's criterion at criticality over the whole domain of aspect ratio. In Fig. 6.5b, however, the modified version of Jung's criterion, exhibits a mismatch in the region of aspect ratios where neck-in influences the stability, and an additional phase δ_a needs to be added, similar to (4.14) for the viscous–inertia model.

Chapter 7
One-dimensional Viscoelastic Models

Abstract This chapter focus on the effect of viscoelastic fluids. For this reason, the following analyses are restricted to a one-dimensional infinite-width film casting configuration and inertia and gravity are neglected. In contrast to Newtonian fluids, there is no unique constitutive equation for describing the correlation between stress and deformation for viscoelastic materials. Therefore, the chapter starts with an overview of typical viscoelastic constitutive equations. After a detailed literature review, viscoelastic film casting models are derived for three different constitutive equations, and the different situation of the initial stress boundary condition as compared to Newtonian fluids is discussed.

7.1 Viscoelastic Constitutive Equations

This section shall give a short introduction to some well-established viscoelastic constitutive equations, with particular focus on elongational deformation, as this is the dominant deformation type in fiber drawing and film casting. The information provided here can be found in several textbooks, e.g., the books of Larson (1988) and Dealy and Larson (2006). The group of viscoelastic constitutive equations can be divided into *integral* and *differential constitutive equations*. The former offer an explicit expression for the stress tensor, which is basically given by an integral over the present and past deformations, which are weighted by a time-dependent memory function. The advantage of this type of constitutive equations is their intuitive, explicit form, which allows a comprehensive physical interpretation and, due to that, the possibility to construct realistic models without additional free parameters. On the other hand, however, the integral form increases the complexity of a model analysis significantly, while differential equations are easier to handle both analytically and numerically. For this reason, many models of viscoelastic material processing employ differential constitutive equations. The benefit of easier analysis, however, is mostly accompanied by the drawback of additional auxiliary variables, which cannot be calculated from molecular theory, but need to be determined by additional material characterization experiments. Within this book, only differential constitutive equations will be studied.

© Springer Nature Switzerland AG 2025

M. Bechert and B. Scheid, *Draw Resonance*, Fluid Mechanics and Its Applications 141,
https://doi.org/10.1007/978-3-031-85568-9_7

The most simple viscoelastic constitutive equation is the *upper-convected Maxwell (UCM) model*, which was already introduced in Sect. 1.5 and is given here again for the sake of completeness,

$$\lambda \overset{\nabla}{\boldsymbol{\tau}} + \boldsymbol{\tau} = 2\,\eta_0\,\mathbf{D}, \tag{7.1}$$

with extra stress tensor $\boldsymbol{\tau}$, rate-of-deformation tensor $\mathbf{D}$ and upper-convected time derivative

$$\overset{\nabla}{\boldsymbol{\tau}} = \partial_t \boldsymbol{\tau} + \mathbf{v} \cdot \nabla \boldsymbol{\tau} - \boldsymbol{\tau} \cdot \nabla \mathbf{v} - (\nabla \mathbf{v})^T \cdot \boldsymbol{\tau}. \tag{7.2}$$

The index '0' refers to the deformation rate dependence of the viscosity, and η_0 is equal to η appearing in the Newtonian constitutive Eq. (1.7). The upper-convected time derivative extends the material time derivative, $D_t := \partial_t + \mathbf{v} \cdot \nabla$, such that the coordinate system follows the deformations of a material element as well. The new parameter appearing in (7.1) is the relaxation time λ, which is a measure for the time scale of elastic processes in the material. Consequently, for deformation times $t \gg \lambda$ only viscous deformation is present. The so-called steady viscosity, however, may still depend on the deformation rate and the relaxation time. In the viscous limit, i.e., $\lambda \to 0$, (7.1) reduces to the Newtonian constitutive equation. The UCM model predicts a constant steady shear viscosity,

$$\eta = \frac{\tau_{xy}}{\dot{\gamma}} = \eta_0, \tag{7.3}$$

independent of the shear rate $\dot{\gamma}$. In contrast to that, the steady elongational viscosity $\mu = \nu/\partial_x u$ in uniaxial and planar deformations depends on the strain rate $\partial_x u$ and the constant Trouton ratio of a Newtonian fluid is only reached in the viscous limit. The expressions can be calculated from (7.1) by setting $D_t \boldsymbol{\tau} = 0$ to

$$\mu = \eta_0 \left(\frac{2}{1 - 2\lambda \partial_x u} + \frac{1}{1 + \lambda \partial_x u} \right), \tag{7.4}$$

for uniaxial deformation, and to

$$\mu = \eta_0 \left(\frac{2}{1 - 2\lambda \partial_x u} + \frac{2}{1 + 2\lambda \partial_x u} \right), \tag{7.5}$$

for planar deformation. This already indicates that the model equations for fiber drawing and film casting differ by more than just a constant factor, like they do for Newtonian fluids. As a consequence, the dynamics and the stability behavior in particular will be different for both processes as well. In order to avoid confusion and to focus on the viscoelastic properties, only the infinite-width film casting model will be analyzed in this chapter. The steps to derive and analyze the fiber drawing model are equivalent and straightforward.

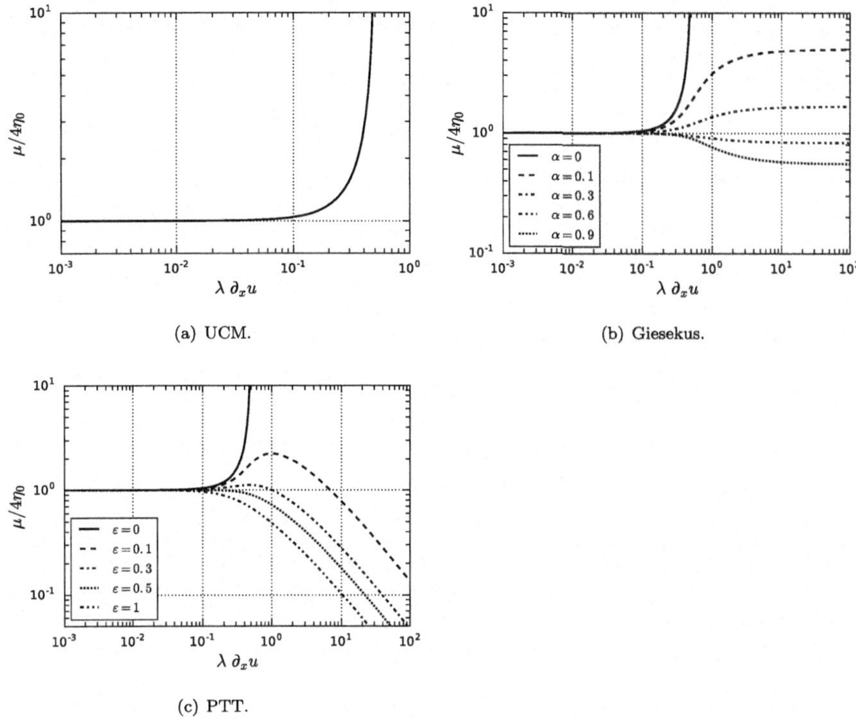

Fig. 7.1 Steady planar elongational viscosity predictions for four different viscoelastic constitutive equations

The planar elongational viscosity (7.5) is plotted in Fig. 7.1a, scaled by the Newtonian value. The UCM model predicts a strong strain hardening effect and for $\partial_x u \to 1/(2\lambda)$, the elongational viscosity diverges. Even though this behavior does not fit to experimental findings for polymer melts, the predictions of UCM process models are often useful, as the steady viscosity and especially the divergence are not always reached within the process time. An empirical way to improve the predictions of the UCM model is to substitute η_0 and λ in (7.1) by functions of the second invariant of the deformation tensor. This extension is called the *White–Metzner (WM) model*. It improves the predictions especially for shear deformations, but is not able to prevent the dramatic strain hardening effect. Two prominent extensions of the UCM model, that predict a finite elongational viscosity, will be presented in the following.

The constitutive equation of the *Giesekus model* reads

$$\lambda \overset{\nabla}{\boldsymbol{\tau}} + \alpha \frac{\lambda}{\eta_0} \boldsymbol{\tau}^2 + \boldsymbol{\tau} = 2\,\eta_0\,\mathbf{D}, \tag{7.6}$$

with nonlinear parameter α. The term in the middle of the left-hand side arises from the following molecular picture. The polymer strands, often modeled by dumbbells, create an anisotropic drag on a single dumbbell due to their mean orientation. The direction of the drag is assumed to be proportional to the anisotropy of the stress tensor, which leads to (7.6). A value of $\alpha = 0$ corresponds to isotropic drag, recovering the UCM model. The anisotropy becomes maximum for $\alpha = 1$. The Giesekus model predicts shear thinning and, depending on α, strain hardening or thinning behavior (see Fig. 7.1b). For large strain rates, a second plateau value of the elongational viscosity is reached, i.e., no strain hardening or thinning for large values of $\lambda \partial_x u$ is predicted. The Giesekus model is said to be well suitable for shear deformations, but not the most appropriate for elongation (Larson, 1988).

The model of *Phan-Thien and Tanner* (PTT) in its simplified form is given by

$$\lambda \overset{\nabla}{\boldsymbol{\tau}} + \exp\left(\frac{\varepsilon\lambda}{\eta_0} tr(\boldsymbol{\tau})\right) \boldsymbol{\tau} = 2\,\eta_0\,\mathbf{D}, \tag{7.7}$$

with nonlinear parameter ε and with $tr(\boldsymbol{\tau})$ denoting the trace of the extra stress tensor. The parameter ε can be seen as a measure for the destruction rate of the polymer network strands. For $\varepsilon > 0$, the excessive strain hardening of the UCM model disappears and the planar elongational viscosity as depicted by Fig. 7.1c shows strain thinning for large strain rates, and, depending on the value of ε, strain hardening for moderate strain rates.

The Giesekus and PTT models both extend the UCM model in a nonlinear way. This means that the difference to the UCM model becomes only visible for rather large deformations, outside of the linear viscoelastic regime, and elongational rheology is typically needed for the determination of the corresponding parameters α or ε. As the two models use only one such nonlinear parameter, they constitute a good compromise between a realistic, finite elongational viscosity and a manageable number of variables for the purpose of a rather qualitative oriented analysis.

7.2 Literature Review

Non-Newtonian effects on draw resonance have been first analyzed by using power-law constitutive equations for the elongational viscosity (see also Sect. 8.3) (Pearson and Shah, 1974; Aird and Yeow, 1983; Van der Hout, 2000). A power-law index larger than unity, i.e., pure strain hardening, increases the critical draw ratio, while a power-law index below unity decreases it as compared to the Newtonian value. In two comprehensive papers, Fisher and Denn (1976, 1977) analyzed the linear stability and the nonlinear dynamics of viscoelastic fiber drawing using an isothermal WM model and a nonisothermal UCM model. Both models predict a sharp increase of the critical draw ratio with increasing relaxation time and reveal an unconditionally stable region for large relaxation times, as well as a second critical draw ratio above which the process becomes stable again. The process parameter space is reduced by an "unattainable" region, where no steady state exist due to high elasticity. The linear stability results of the UCM model will be reviewed as well in Sect. 8.1.1.

Up to now, various viscoelastic constitutive equations have been studied in the scope of draw resonance, among them being models of WM type (Gupta et al., 1996; Lee et al., 2001a), one-dimensional (Silagy et al., 1996) and two-dimensional (Silagy et al., 1998; Kim et al., 2005) UCM film casting models, PTT models for fiber drawing (Lee et al., 2001a) as well as for one-dimensional (Lee et al., 2007) and two-dimensional (Shin et al., 2007) film casting, and a study of the so-called *eXtended Pom-Pom (XPP) model* (van der Walt et al., 2012). German and Khayat (2008) investigated the combined effect of inertia and elasticity in film casting of a polymer solution by employing a PTT constitutive model. In two associated papers, Iyengar and Co (1993, 1996) analyzed the steady state and linear stability of film casting of a modified Giesekus fluid.[1] They discussed their results in correlation with the steady shear and uniaxial elongational viscosities and their dependence on the deformation rates, and stated that a thinning behavior leads to destabilization, while strain hardening may lead to an increase of the critical draw ratio, provided that the relaxation time and the strain rates are in the appropriate ranges.

Zavinska et al. (2008) (see also the thesis of Zavinska (2008)) also utilized a modified version of the Giesekus model to predict the critical draw ratio in film casting. Their results, however, are questionable as they do not converge to the Newtonian limit for small relaxation times. Smith and Stolle (2000b) used a material description of viscoelastic film casting and a constitutive equation which is described as an equivalent to the UCM model. In contrast to the strong stabilizing effect for the UCM model, however, they reported a destabilizing effect for increasing relaxation time. Lately, Gupta and Chokshi (2015) presented a weakly nonlinear stability analysis of fiber drawing using the XPP model. Combined with linear stability results, they found a second transition from supercritical to subcritical stability behavior. Later Gupta et al. (2016) extended this study to cover the influence of swelling after extrusion through the die. Most of these studies reveal that purely strain thinning materials show a destabilization with increasing elasticity and that strain hardening may lead to a stability maximum. Only exceptions are the UCM and WM models, as already mentioned above.

Apart from draw resonance studies, there is a group of steady state investigations accounting for the fact that viscoelastic materials in general have not only a single, but a continuous spectrum of relaxation time. The continuous spectrum, however, is in most cases approximated by a discrete multimode spectrum. Exceptions are the two papers of Fulchiron et al. (1995a, b), who use a postulated function for the relaxation time spectrum and an integral constitutive equation to investigate the influence of the spectrum width on the steady states of fiber drawing. For both

[1] The modification introduces a further parameter accounting for a finite polymer chain extensibility, which in turn leads to strain thinning at large strain rates.

fiber drawing (Papanastasiou et al., 1987; Chen and Papanastosiou, 1987; Fulchiron et al., 1997; Mitsoulis and Beaulne, 2000) and film casting (Alaie and Papanastasiou, 1991; Beaulne and Mitsoulis, 1999), steady state solutions for integral constitutive equations have been analyzed, accounting for a shear pre-history in the die and/or a broad distribution of relaxation times. Satoh et al. (2001) presented an analysis of a Larson model with a discrete number of relaxation times. Christodoulou et al. (2000) and Dhadwal et al. (2017) have analyzed the linear stability of film casting of a multimode PTT model.

7.3 Viscoelastic Film Models

7.3.1 *Model Equations*

The derivation of the model equations for viscoelastic film casting is similar to the one for Newtonian fluids presented in Sect. 2.2. Nevertheless, due to the different form of the constitutive equations, the procedure is not straightforward and will be presented in the following. First, the equations corresponding to the UCM model will be deduced. Afterwards, the additional terms of the Giesekus and PTT models will be handled. As already mentioned above, gravity, inertia and finite-width effects will be neglected within this chapter.

While the stress tensor for Newtonian fluids is given explicitly by the constitutive equation, (7.1), (7.6), and (7.7) yield an implicit differential definition. Applying the scaling rules (2.28) makes it possible to write the single components of (7.1) as

$$\begin{aligned} &De\left(\left(\partial_t\tau_{xx} + v\partial_y\tau_{xx} + u\partial_x\tau_{xx}\right)\beta^2 \right.\\ &\left.-2\left(\beta^2\partial_x u\tau_{xx} + \partial_y u\tau^*_{xy}\right)\right) + \beta^2\tau_{xx} = 2\beta^2\partial_x u, \end{aligned} \tag{7.8a}$$

$$De\left(\partial_t\tau_{yy} + v\partial_y\tau_{yy} + u\partial_x\tau_{yy} - 2\left(\partial_y v\tau_{yy} + \partial_x v\tau^*_{xy}\right)\right) + \tau_{yy} = 2\partial_y v, \tag{7.8b}$$

$$\begin{aligned} &De\left(\partial_t\tau^*_{xy} + v\partial_y\tau^*_{xy} + u\partial_x\tau^*_{xy} - \partial_y v\tau^*_{xy}\right.\\ &\left.-\beta^2\partial_x v\tau_{xx} - \partial_x u\tau^*_{xy} - \partial_y u\tau_{yy}\right) + \tau^*_{xy} = \beta^2\partial_x v + \partial_y u, \end{aligned} \tag{7.8c}$$

with the Deborah number De, defined as

$$De = \lambda\frac{u_0}{L}, \tag{7.9}$$

as dimensionless control parameter and $\tau^*_{xy} = \beta\tau_{xy}$. Analogue to Sect. 2.2, the momentum equation and the stress boundary condition lead to $\sigma^{*(0)}_{xy} = \tau^{*(0)}_{xy} = 0$. The leading order of (7.8c) thus becomes

$$De\ \partial_y u^{(0)}\tau^{(0)}_{yy} = -\partial_y u^{(0)}. \tag{7.10}$$

Assuming that $\partial_y u^{(0)}$ is non-zero would lead to

$$\tau_{yy}^{(0)} = -\frac{1}{De}, \tag{7.11}$$

which predicts a diverging transversal stress component in the Newtonian limit, $De \to 0$. As this is incorrect, the axial velocity component must not vary along the thickness at leading order, i.e., $\partial_y u^{(0)} = 0$. The derivation of the integrated continuity (2.33) is then straightforward.

Following the steps in Sect. 2.2, $\sigma_{xx}^{(0)}$ is equal to the normal stress difference $\nu^{(0)} = \tau_{xx}^{(0)} - \tau_{yy}^{(0)}$, independent of the constitutive equation. In contrast to the Newtonian case, however, the normal stress difference is not independent of y in general, and integrating the momentum equation over the film thickness leads to

$$\partial_x \left(h\, \overline{\nu}^{(0)}\right) = 0, \tag{7.12}$$

with the depth-averaged normal stress difference

$$\overline{\nu}^{(0)} = \frac{1}{h}\int_{-h/2}^{h/2} dy\; \nu^{(0)} = \overline{\tau}_{xx}^{(0)} - \overline{\tau}_{yy}^{(0)}. \tag{7.13}$$

Integrating the first non-vanishing order of (7.8a) and (7.8b) over the thickness, utilizing the continuity equation, finally yields

$$De\left(\partial_t \overline{\nu}^{(0)} + u^{(0)}\partial_x \overline{\nu}^{(0)} + 2\partial_x u^{(0)}\left(\overline{\nu}^{(0)} - 2\overline{\tau}_{xx}^{(0)}\right)\right) + \overline{\nu}^{(0)} = 4\partial_x u^{(0)}, \tag{7.14a}$$

$$De\left(\partial_t \overline{\tau}_{xx}^{(0)} + u^{(0)}\partial_x \overline{\tau}_{xx}^{(0)} - 2\partial_x u^{(0)}\overline{\tau}_{xx}^{(0)}\right) + \overline{\tau}_{xx}^{(0)} = 2\partial_x u^{(0)}. \tag{7.14b}$$

As before, the superscripts "(0)" will be omitted from now on, as well as the bars. The model Eqs. (2.33), (7.12) and (7.14) have the same mathematical form like the ones used in literature, e.g., by Fisher and Denn (1976). Nevertheless it is important to note that the membrane approximation, i.e., $\beta \ll 1$, does not imply that the stress tensor becomes independent of y, and that the extra stress tensor components appearing in the model equations are averaged values. This apparent subtlety becomes important especially for the nonlinear extensions appearing in the Giesekus and PTT models, as shown next.

The Giesekus model (7.6) adds a quadratic stress term to the constitutive equation of the UCM model. The scaled xy-component of this term is $\alpha\, De\, \tau_{xy}^{*}\left(\tau_{xx} + \tau_{yy}\right)$, which becomes zero at leading order and therefore $\partial_y u = 0$ for the Giesekus model as well. The only remaining parts contributing to the first non-vanishing orders of the xx- and yy-components are $\alpha De\left(\tau_{xx}^{(0)}\right)^2$ and $\alpha De\left(\tau_{yy}^{(0)}\right)^2$, respectively. In order to integrate over the thickness, it is necessary to assume additionally that the y-dependence of $\tau_{xx}^{(0)}$ and $\tau_{yy}^{(0)}$ is proportional to $\mathcal{O}(\beta)$, i.e.,

$$|\tau_{ii}^{(0)} - \overline{\tau}_{ii}^{(0)}| = \mathcal{O}(\beta), \tag{7.15}$$

so that in particular

$$\frac{1}{h}\int_{-h/2}^{h/2} dy\, \left(\tau_{ii}^{(0)}\right)^2 = \left(\overline{\tau}_{ii}^{(0)}\right)^2, \tag{7.16}$$

for $i \in \{x, y\}$. The averaged Giesekus constitutive equations then become

$$De\left(\partial_t \overline{\nu}^{(0)} + u^{(0)}\partial_x \overline{\nu}^{(0)} + \left(2\partial_x u^{(0)} - \alpha\overline{\nu}^{(0)}\right)\left(\overline{\nu}^{(0)} - 2\overline{\tau}_{xx}^{(0)}\right)\right) + \overline{\nu}^{(0)} = 4\partial_x u^{(0)}, \tag{7.17a}$$

$$De\left(\partial_t \overline{\tau}_{xx}^{(0)} + u^{(0)}\partial_x \overline{\tau}_{xx}^{(0)} - 2\partial_x u^{(0)}\overline{\tau}_{xx}^{(0)} + \alpha\left(\overline{\tau}_{xx}^{(0)}\right)^2\right) + \overline{\tau}_{xx}^{(0)} = 2\partial_x u^{(0)}. \tag{7.17b}$$

Finally, the exponential factor introduced by the PTT model does not affect the deduction of (7.10), and the axial velocity does not change along the thickness for the PTT model as well. Assumption (7.15) also implies that

$$\frac{1}{h}\int_{-h/2}^{h/2} dy\, \mathrm{e}^{\varepsilon De\left(\tau_{xx}^{(0)}+\tau_{yy}^{(0)}\right)}\tau_{ii}^{(0)} = \mathrm{e}^{\varepsilon De\left(\overline{\tau}_{xx}^{(0)}+\overline{\tau}_{yy}^{(0)}\right)}\overline{\tau}_{ii}^{(0)}, \tag{7.18}$$

and the depth-averaged constitutive equations for the PTT model can be written as

$$De\left(\partial_t \overline{\nu}^{(0)} + u^{(0)}\partial_x \overline{\nu}^{(0)} + 2\partial_x u^{(0)}\left(\overline{\nu}^{(0)} - 2\overline{\tau}_{xx}^{(0)}\right)\right) + \mathrm{e}^{\varepsilon De\left(2\overline{\tau}_{xx}^{(0)}-\overline{\nu}^{(0)}\right)}\overline{\nu}^{(0)} = 4\partial_x u^{(0)}, \tag{7.19a}$$

$$De\left(\partial_t \overline{\tau}_{xx}^{(0)} + u^{(0)}\partial_x \overline{\tau}_{xx}^{(0)} - 2\partial_x u^{(0)}\overline{\tau}_{xx}^{(0)}\right) + \mathrm{e}^{\varepsilon De\left(2\overline{\tau}_{xx}^{(0)}-\overline{\nu}^{(0)}\right)}\overline{\tau}_{xx}^{(0)} = 2\partial_x u^{(0)}. \tag{7.19b}$$

Note that the superscripts "(0)" will be omitted in the subsequent chapters.

7.3.2 *Initial Stress Boundary Condition*

A drawback of using differential constitutive equations instead of integral ones is the necessity to specify an additional boundary condition for the stress tensor components, as it is not possible to reduce the unknowns to one single stress component. This boundary condition is a priori unknown and not implicitly given by the process–in fact it physically overspecifies the system, but is mathematically needed. It cannot be chosen arbitrarily, though. Typically, the ratio of the transversal stress τ_{yy} to the normal stress difference ν at the inlet is prescribed. Many authors cited in Sect. 7.2 define this ratio to be very small, following the argument that the transversal stress vanishes due to complete relaxation at the point of largest swelling, which is taken as the starting point for modeling. As pointed out by Papanastasiou et al. (1996), this assumption does not fit to the Newtonian limit, where the transversal stress is proportional to $\partial_x u$, and thus cannot be correct for the entire range of relaxation times. To overcome this problem, Papanastasiou et al. (1996) implemented a so-called free stress boundary condition method for the steady fiber drawing process. Using finite element methods, they refine the stress inlet condition so that it is specified by the condition given implicitly by the constitutive equation itself. For the UCM model they use for demonstration, this method yields smooth solutions for the stress tensor components, without any boundary effects close to the inlet. Additionally, they showed that the calculated initial stress value can be used as boundary condition for a common initial value approach, which leads to identical solutions. This method was also partly used by van der Walt et al. (2012) for their analysis of the XPP model, but yielding useful values only for parts of the parameter space.

The method of numerical continuation, which is used within this book to calculate the steady state and perturbation solutions, cannot be changed straightforwardly to apply the free stress boundary condition method described above. Nevertheless, as the continuation starts at $De = 0$, it is inevitable to specify a stress boundary condition which fulfills the Newtonian limit. Based on the data presented by Papanastasiou et al. (1996) and the argument that the transversal stress should be zero at the inlet for high elasticity, we will use the following empirical inlet stress boundary condition,

$$\left.\frac{\tau_{yy}}{\nu}\right|_{x=0} = \left.\frac{\tau_{xx}}{\nu}\right|_{x=0} - 1 = -\frac{1}{2}\exp\left(-2De^{0.7}\right), \tag{7.20}$$

in which (7.13) has been used. The exponential decay of τ_{yy} with increasing Deborah number is a purely empirical assumption, but still reasonable due to the arguments above. For $De = 0$, the Newtonian value of $-1/2$ is recovered. The coefficient of 2 and the exponent of 0.7 is chosen for reasons which will become clear in Sect. 8.2, as the boundary effect caused by the inexactly chosen inlet stress condition is almost completely suppressed in the effective elongational viscosity of the PTT model. The critical draw ratio can generally change with different initial stress conditions. The qualitative shape of the neutral stability curve usually remains the same, though, and it is therefore important for a comparison of the various models that the same condition is used in all cases.

Note that a less empirical solution of the initial stress boundary condition exists, which actually reflects the stress-free boundary condition. However, as the essential results are unaffected by this refinement, and as the empirical condition is easier to apply in many cases, we still prefer to use it in this book and refer the interested reader to Bechert (2020) for further details.

Chapter 8
Influence of Viscoelasticity

Abstract The focus of this chapter lies on the influence of viscoelasticity on draw resonance. A linear stability analysis of all viscoelastic models considered in the previous chapter is presented, examining the influence of the Deborah number on the critical draw ratio, followed by a direct comparison and evaluation of the constitutive equations with respect to their stability predictions. The last section covers studies on the viscoelastic stability mechanism, including analyses of generalized Newtonian fluids and the influences of strain hardening and thinning effects.

8.1 Linear Stability Analysis

In this section, the dependence of the critical draw ratio on the Deborah number and possible nonlinear parameters as obtained by linear stability analysis is presented and discussed for the previously presented UCM, Giesekus, and PTT film casting models. The procedure is identical to the one described in Sect. 3.2, with the following additional ansatz for the xx-component of the extra stress tensor,

$$\tau_{xx}(x,t) = t_s(x)\left(1 + T(x)\,\mathrm{e}^{\omega t}\right) \quad \text{with } |T| \ll 1\,, \tag{8.1}$$

where t_s is the steady state and T the complex perturbation. The main purpose of this section is to point out the main differences in the stability predictions of the different constitutive equations, as well as their suitability for further modeling.

8.1.1 UCM Model

The steady state equations of the UCM film casting model, i.e., (2.33), (7.12) and (7.14), are given by

© Springer Nature Switzerland AG 2025

M. Bechert and B. Scheid, *Draw Resonance*, Fluid Mechanics and Its Applications 141,
https://doi.org/10.1007/978-3-031-85568-9_8

$$u_s' = \frac{\nu_s}{4c}, \tag{8.2a}$$

$$h_s' = -\frac{h_s \nu_s}{4u_s c}, \tag{8.2b}$$

$$\nu_s' = \frac{\nu_s^2}{4u_s c}, \tag{8.2c}$$

$$t_s' = \frac{1}{De\, u_s}\left(2u_s'\,(1 + De\, t_s) - t_s\right), \tag{8.2d}$$

with

$$c = 1 + De\left(t_s - \frac{3}{4}\nu_s\right). \tag{8.3}$$

The perturbations are governed by the following equations,

$$U' = \frac{\nu_s}{4u_s c^2}\left[(1 + De\,(t_s + \omega c))\,N - c\,U + De\,(\omega c H - t_s T)\right], \tag{8.4a}$$

$$H' = -\left(\frac{\omega}{u_s} H + U'\right), \tag{8.4b}$$

$$N' = -H', \tag{8.4c}$$

$$T' = \frac{1}{De\, u_s t_s}\left[t_s U + 2u_s(1 + De\, t_s)U' - \left(\omega De\, t_s + \frac{\nu_s}{2c}\right) T\right]. \tag{8.4d}$$

The system of Eqs. (8.2) and (8.4) are solved by continuation with the boudnary conditions (3.5), (3.11c), (3.28) and (7.20), which implies $T(0) = 0$. The resulting neutral stability curves are shown in Fig. 8.1. The data calculated with numerical continuation using AUTO-07P coincides perfectly with the one presented by Silagy et al. (1996). As already mentioned in Sect. 7.1, fiber drawing governed by uniaxial deformation will result in a different stability curve due to the differences in the constitutive equations. For this reason, the result of Fisher and Denn (1976) has only qualitative similarities. Denn et al. (1975) revealed that steady state solutions are only possible for draw ratios $Dr \leq 1 + De^{-1}$, where the equality holds if the time within which a fluid particle flows from inlet to outlet is approximately equal to the relaxation time. This is a consequence of the strong stresses predicted by the UCM model for large relaxation times and strain rates. The resulting "unattainable" region is shaded in gray in Fig. 8.1.

The UCM model predicts, apart from a practically negligible stability minimum shown by the inset of Fig. 8.1, a purely stabilizing influence of the Deborah number, i.e., increasing elastic fluid properties. At a critical value $De_c = 0.013$, the stability curve shows a folding point and as a result, a second critical draw ratio, above which the process becomes stable again, existing for $De < De_c$. On the other hand, the system becomes unconditionally stable for $De > De_c$. As noted by Co (Hatzikiriakos and Migler, 2004), the second critical draw ratio has never been observed experimentally. For these reasons, the UCM model should be utilized merely in the scope of general, theoretical considerations or as limit case of the other models. Steady states and perturbations at $De = 10^{-2}$ for the lower and upper branches of the stability curve are shown in Fig. 8.2.

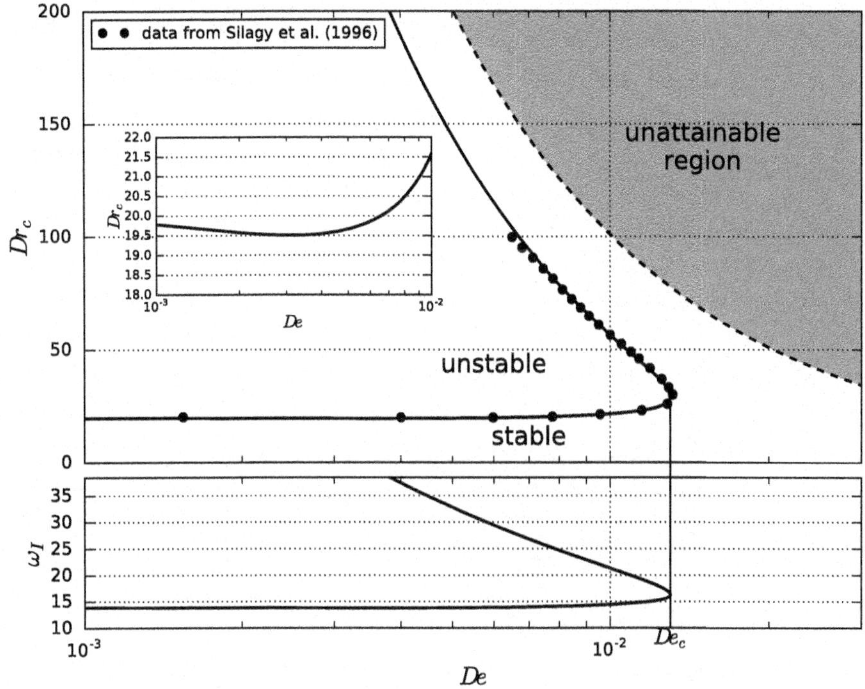

Fig. 8.1 Neutral stability curve of the infinite-width UCM film casting model. The calculation of the critical draw ratio with AUTO-07P shows perfect coincidence with the curve presented by Silagy et al. (1996). The shaded region corresponds to parameter combinations for which no steady state solution exists due to elastic effects, as shown by Denn et al. (1975). For $De > De_c = 0.013$, no critical draw ratio exist, i.e., the system is unconditionally stable. The inset visualizes the small stability minimum located at $De = 3.1 \times 10^{-3}$

8.1.2 Giesekus Model

The steady state equations of the Giesekus model, i.e., (2.33), (7.12) and (7.17), are very similar to the UCM model equations and given by

$$u_s' = \frac{\nu_s c_G}{4c} , \tag{8.5a}$$

$$h_s' = -\frac{h_s \nu_s c_G}{4u_s c} , \tag{8.5b}$$

$$\nu_s' = \frac{\nu_s^2 c_G}{4u_s c} , \tag{8.5c}$$

$$t_s' = \frac{1}{De\, u_s} \left(2u_s' \left(1 + De\, t_s\right) - t_s \left(1 + \alpha\, De\, t_s\right)\right) , \tag{8.5d}$$

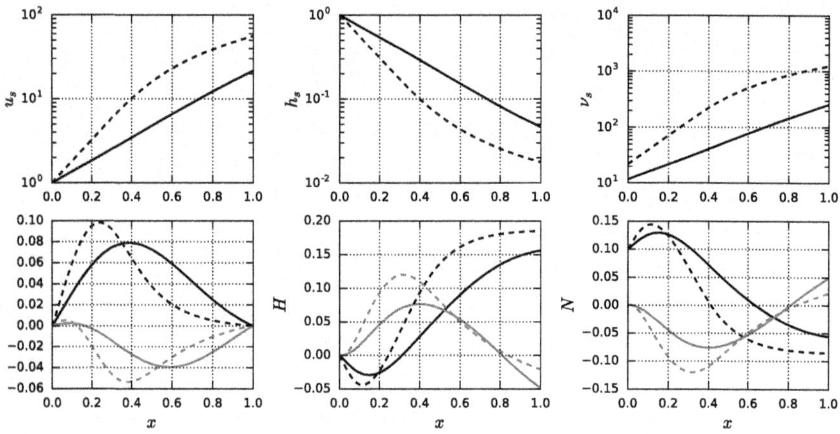

Fig. 8.2 Steady states and perturbations at criticality for the infinite-width UCM film casting model at $De = 10^{-2}$. The solid lines correspond to the lower branch of the neutral stability curve, where $Dr_c = 21.58$ and the dashed lines to the upper branch, where $Dr_c = 56.45$. The real parts of the complex perturbations are plotted in black and the imaginary parts in gray

with

$$c_G = 1 + \alpha\, De\,(2t_s - v_s)\;, \tag{8.6}$$

and c as defined by (8.3). The perturbation equations for U and T can be written as

$$\begin{aligned} U' =& \frac{v_s}{4u_s c^2}\left[\left(1 + De\left(t_s + \omega c + 2\alpha\left(c\,t_s - v_s\left(1 + \frac{De}{4}\left(t_s - \frac{3}{2}v_s\right)\right)\right)\right)\right)N\right.\\ &\left. - c\,c_G U + De\left(\omega c H - t_s\left(1 + \alpha\left(De\,\frac{v_s}{2} - 2\right)\right)T\right)\right]\,, \end{aligned} \tag{8.7a}$$

$$\begin{aligned} T' =& \frac{1}{De\,u_s t_s}\Big[t_s U\,(1 + \alpha\,De\,t_s) + 2u_s(1 + De\,t_s)U'\\ &-\left(De\,t_s\,(\omega + \alpha\,t_s) + \frac{v_s c_G}{2c}\right)T\Big]\,, \end{aligned} \tag{8.7b}$$

while (8.4b) and (8.4c) remain unchanged.

Figure 8.3 shows the resulting neutral stability curves obtained by numerical continuation with respect to De for various values of α. For most values of α, increasing the Deborah number has primarily a destabilizing effect, but a stability maximum may appear for small values of α, as visible here for $\alpha = 0.1$, where the maximum is located at $De = 0.042$. Increasing α has a destabilizing effect for most values of De. Exceptions may occur for values of α close to unity, e.g., the critical draw ratio for $\alpha = 0.9$ is above the one for $\alpha = 0.6$ for a particular range of De. However, the results for $\alpha = 0.9$ may be incorrect, as the first perturbation mode,

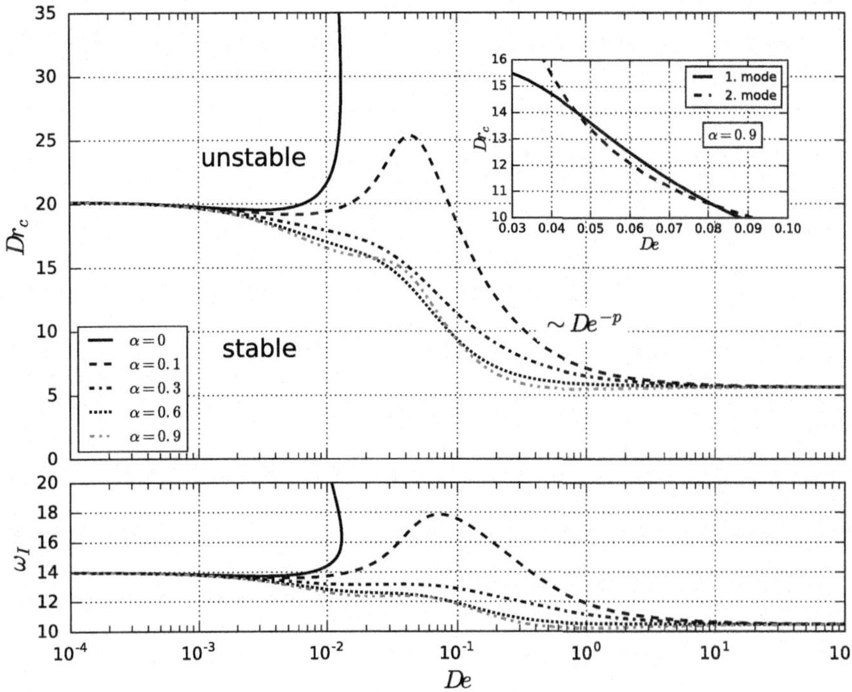

Fig. 8.3 Neutral stability curves of the infinite-width Giesekus film casting model for various values of the nonlinear parameter α. For $De > 0.1$, the critical draw ratio decreases exponentially with De. For $\alpha = 0.9$, the critical draw ratio corresponding to the first draw resonance mode becomes larger than the one corresponding to the second mode within a particular range of De, as shown by the inset. For this reason, the curve is not correct for all values of De and is plotted in gray

i.e., the one with the smallest imaginary part of the eigenvalue, does not appear to be the most unstable for all Deborah numbers, which is visualized by the inset of Fig. 8.3.

For this reason, the neutral stability curves as obtained by tracking the first perturbation mode with AUTO-07P are compared with the ones resulting from a full eigenvalue spectrum analysis. For several values of De, the draw ratio is increased stepwise and the spectrum is calculated as described in Appendix A, until the largest real part of the eigenvalues becomes equal or above zero. The step size for Dr is set to 0.1, and the arithmetic mean value of the largest draw ratio with purely negative eigenvalue real parts and the smallest draw ratio with at least one real part equal or above zero is taken as Dr_c, with an error of ± 0.05.

The results of both numerical methods are visualized in Fig. 8.4 for $\alpha = 0.9$. For most values of De, both methods coincide perfectly, but within $0.04 < De < 0.1$, the eigenvalue analysis leads to slightly smaller critical draw ratios than the numerical continuation does. The eigenvalue spectra shown by the insets of Fig. 8.4 reveal that within this range, the imaginary part of the eigenvalues does not decrease monotoni-

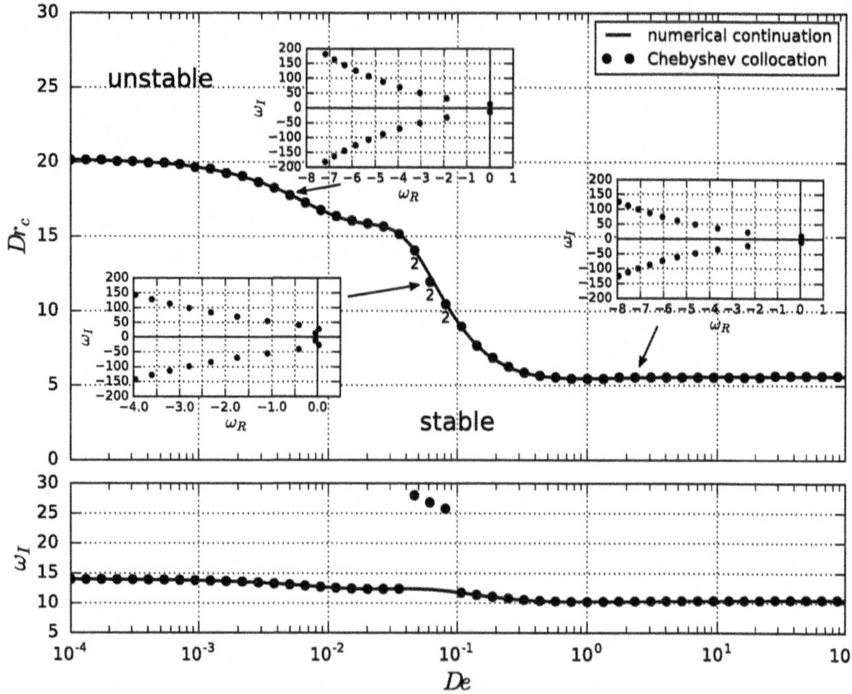

Fig. 8.4 Comparison of the Giesekus neutral stability curves for $\alpha = 0.9$ as calculated by numerical continuation and by eigenvalue spectrum analysis via Chebyshev collocation. The number below the points indicates the number of the leading perturbation mode and is omitted for the first mode. The spectra at three different Deborah numbers are shown by insets

cally with increasing real part. As a consequence, the second mode becomes the most unstable mode, which is indicated as well by the label '2' below the corresponding data points. The variation of the critical draw ratio, however, remains small, and it can be seen as well in the spectrum plot that the real parts of the first and second modes are close together. The comparative analysis has been performed for $\alpha = 0.1$ to 0.9 with an increment step size of 0.1 and the effect of leading mode switching was observed only for $\alpha = 0.9$. Figure 8.5 shows exemplarily the high coincidence of both analysis methods for $\alpha = 0.1$ and 0.6. For this reason, the influence of higher perturbation modes on the critical draw ratio can be considered as small and negligible for the Giesekus model. The dominating oscillation frequency, however, which is determined by the imaginary part of the eigenvalue, changes noncontinuously if the leading mode changes, which is shown at the bottom of Fig. 8.4.

As the critical draw ratio curves in Fig. 8.3 have been proven to be correct, with only a small error for $\alpha = 0.9$, the intersection of the curves for $\alpha = 0.6$ and $\alpha = 0.9$ is also correct, leading to a nonmonotonous stability dependence on α for $De = \mathcal{O}\left(10^{-2}\right)$. It has to be noted that, as already mentioned above, the selection of the initial stress boundary condition (7.20) influences both the critical draw ratio

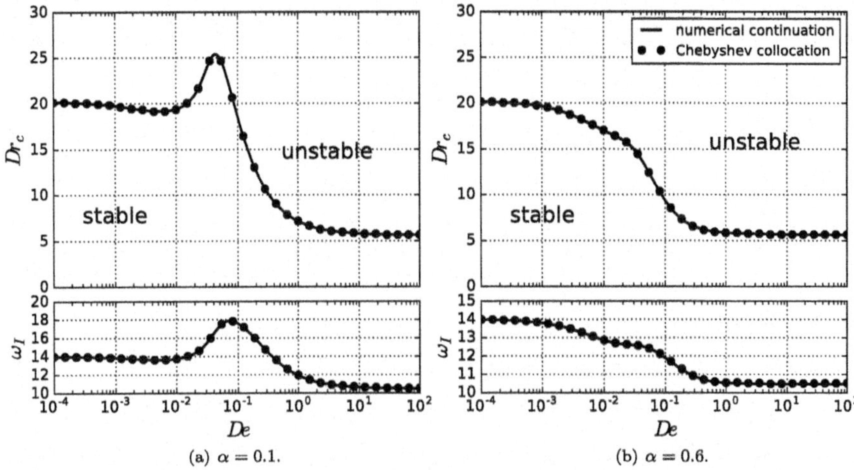

Fig. 8.5 Comparison of the Giesekus neutral stability curves for $\alpha = 0.1$ and 0.6 as calculated by numerical continuation and by eigenvalue spectrum analysis via Chebyshev collocation. The first perturbation mode is the most unstable one for the entire range of De in both cases

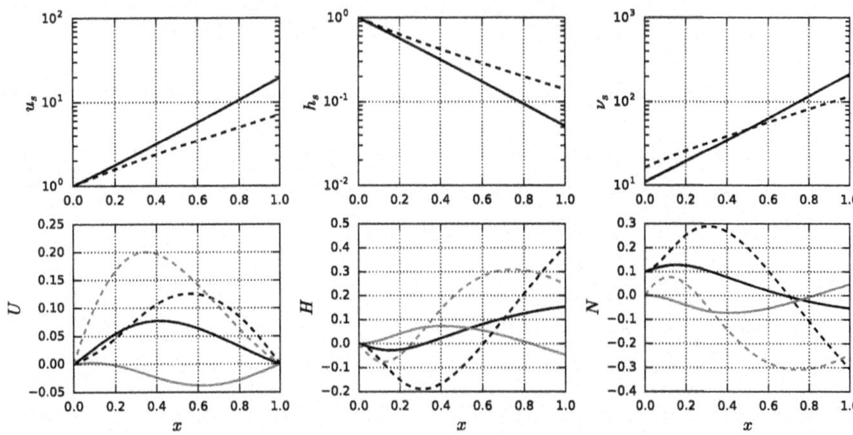

Fig. 8.6 Steady states and perturbations at criticality for the infinite-width Giesekus film casting model with $\alpha = 0.1$. The solid lines correspond to $De = 10^{-2}$ and the dashed lines to $De = 1$. The real parts of the complex perturbations are plotted in black and the imaginary parts in gray

and the eigenvalue spectrum, which means that for other boundary conditions, the effect of mode switching may be more influencing than here.[1] For $\alpha = 0.1$, Fig. 8.6 visualizes steady states and perturbations for $De = 10^{-2}$ and 1.

The neutral stability curves shown in Fig. 8.3 can be fitted for $De > 0.1$ with a shifted power law,

[1] The nonmonotonous stability dependence on α may also disappear or become more pronounced for other initial stress boundary conditions.

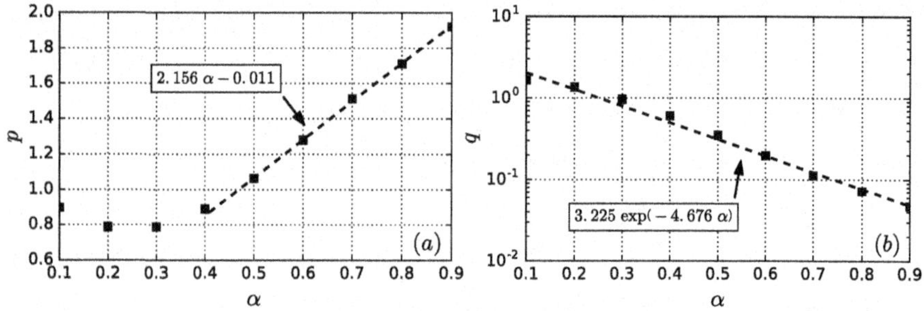

Fig. 8.7 Dependence of the fit parameters p and q of (8.8) on α

$$Dr_{\mathrm{c}}|_{De>0.1} = \frac{q}{De^{p}} + Dr_{\mathrm{c},\infty} \,, \tag{8.8}$$

with free parameters p, q and $Dr_{\mathrm{c},\infty}$. For $\alpha = 0.9$, the data resulting from the eigenvalue spectrum analysis was utilized. For all values of α, R^2_{adj} lies above 0.98, indicating the high quality of the fits. The plateau value $Dr_{\mathrm{c},\infty}$ was found to be constant for all values of α and determined as 5.60 ± 0.04. Figure 8.7 visualizes the correlation between α and the parameters p and q. While the critical exponent p does not vary much for $\alpha < 0.4$, with a value of approximately 0.85, it increases linearly for larger values of α. A possible linear fit is given in Fig. 8.7a. The dependence of q on α can be approximated by a decaying exponential function, which itself is given in Fig. 8.7b. Comparing these results with the ones of the Larson model (not shown) reveals that while the critical draw ratio for large Deborah numbers is higher for the Giesekus model, the parameters p and q are in the same range.

8.1.3 PTT Model

Utilizing the PTT constitutive Eq. (7.19) together with (2.33) and (7.12) leads to the following steady state equations,

$$u_s' = \frac{\nu_s}{4\,c\,c_{\mathrm{PTT}}} \,, \tag{8.9a}$$

$$h_s' = -\frac{h_s \nu_s}{4 u_s c\, c_{\mathrm{PTT}}} \,, \tag{8.9b}$$

$$\nu_s' = \frac{\nu_s^2}{4 u_s c\, c_{\mathrm{PTT}}} \,, \tag{8.9c}$$

$$t_s' = \frac{1}{De\, u_s}\left(2u_s' \left(1 + De\, t_s\right) - \frac{t_s}{c_{\mathrm{PTT}}}\right) \,, \tag{8.9d}$$

with

$$c_{\mathrm{PTT}} = \mathrm{e}^{\varepsilon De(\nu_s - 2t_s)} , \tag{8.10}$$

and c as defined by (8.3). The equations for the perturbations U and T are given by

$$\begin{aligned} U' =& \frac{\nu_s}{4u_s c^2 c_{\mathrm{PTT}}} [(1 - De\,(c\,(\varepsilon \nu_s - \omega c_{\mathrm{PTT}}) - t_s))\,N \\ & - c\,U + De\,(\omega\, c\, c_{\mathrm{PTT}} H - t_s\,(1 - 2\,\varepsilon\, c)\,T)] , \end{aligned} \tag{8.11a}$$

$$\begin{aligned} T' =& \frac{1}{De\, u_s t_s c_{\mathrm{PTT}}} \Big[t_s U + \varepsilon De\, t_s \nu_s N + 2u_s c_{\mathrm{PTT}}(1 + De\, t_s)U' \\ & - \left(De\, \omega t_s c_{\mathrm{PTT}} + \frac{\nu_s}{2c} + 2\,\varepsilon De\, t_s^2 \right) T \Big] , \end{aligned} \tag{8.11b}$$

and (8.4b) and (8.4c) remain unchanged. The neutral stability curves with respect to the Deborah number as calculated by numerical continuation are depicted in Fig. 8.8. The qualitative predictions for the critical draw ratio are comparable to the ones resulting from the Giesekus model. While the UCM curve is obtained for $\varepsilon = 0$, increasing the Deborah number has a mainly destabilizing effect for other values of ε. For low values of ε a small stability maximum appears, e.g., located at $De = 0.271$ for $\varepsilon = 0.1$. For $De > 0.1$, the critical draw ratio decreases exponentially with De.

The first perturbation mode of the linear stability analysis is not necessarily the most unstable one for all values of ε, as it was already observed for the Giesekus model. The inset of Fig. 8.8 shows this behavior for $\varepsilon = 1$. The neutral stability curves with respect to De have therefore been evaluated as well by using the previously described eigenvalue spectrum analysis for values of ε from 0.1 to 1 with increment step size of 0.1. A comparison of both methods for $\varepsilon = 1$ is shown in Fig. 8.9. In contrast to the Giesekus model (see Fig. 8.4), the stability curve obtained by numerical continuation clearly differs from the one resulting from the eigenvalue spectrum analysis for $0.02 < De < 1$. The insets, which visualize the eigenvalue spectra at three different Deborah numbers, reveal that for rather small values of De, the first mode is definitely the most unstable, but for moderate values of De, higher perturbation modes[2] have a larger real part than the first mode and dominate the stability behavior. Consequently, the dimensionless oscillation frequency at criticality ω_I exhibits several discontinuities. For $De > 1$, both analysis methods coincide again, but the real parts of the first two perturbation modes are still close together. The phenomenon of perturbation mode switching was observed for $\varepsilon \geq 0.5$. For $\varepsilon = 0.5$, the error in the critical draw ratio of the continuation method is still negligible, but distinct differences in the oscillation frequency are present. Figure 8.10 shows exemplarily the comparison of both analysis methods for $\varepsilon = 0.1$ and 0.5.

---

[2] Higher modes than the fifth have been observed as well to be the most unstable in some cases.

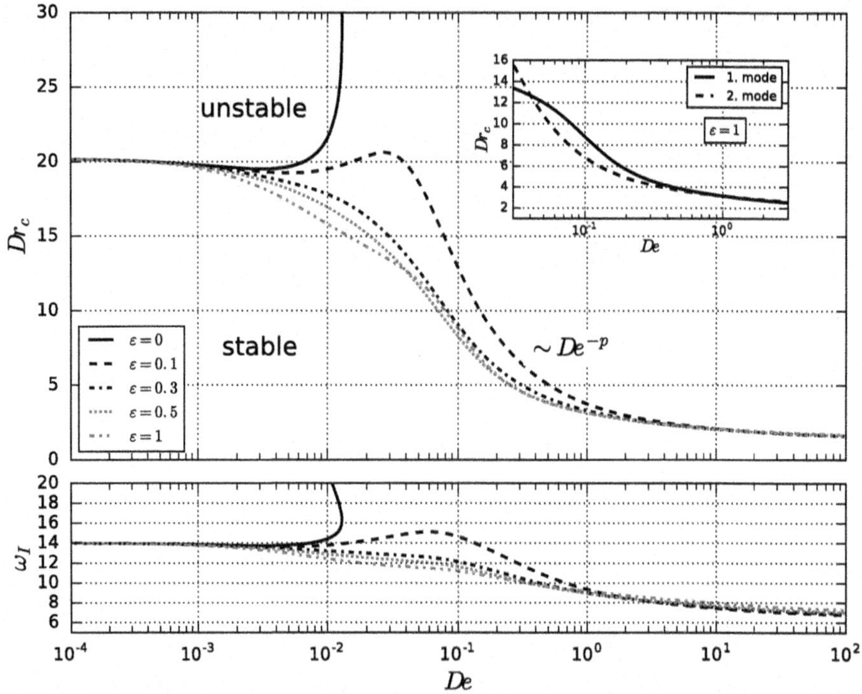

Fig. 8.8 Neutral stability curves of the infinite-width PTT film casting model for various values of the nonlinear parameter ε. For $De > 0.1$, the critical draw ratio decreases exponentially with De. For ε equal to 0.5 and 1, the critical draw ratio corresponding to the first draw resonance mode becomes larger than the one corresponding to the second mode within a particular range of De, as shown by the inset for $\varepsilon = 1$. For this reason, the curves are not correct for all values of De and plotted in gray

Figure 8.8 shows that the partly incorrect, gray plotted stability curves for $\varepsilon = 0.5$ and 1 intersect. The eigenvalue spectrum analysis reveals, however, that this intersection is caused by the switching of perturbation modes. The correct stability curves do not intersect, which means that increasing the nonlinear parameter ε has a purely destabilizing effect for the entire range of Deborah numbers. Figure 8.11 depicts steady states and perturbations at criticality for $\varepsilon = 0.1$ at $De = 10^{-2}$ and 1.

The behavior of the stability curves for $De > 0.1$ was analyzed by fitting with the power-law (8.8), analogue to the previous case of the Giesekus models. The stability curves obtained by the eigenvalue spectrum analysis were used in all cases and R^2_{adj} is always above 0.99. The plateau value for large Deborah numbers was determined to be constant, with a value of $Dr_{c,\infty} = 1.70 \pm 0.02$. This value is smaller than the one obtained from the Giesekus model. A possible reason for this is the different behavior of the predicted elongational viscosities for large values of $De\, \partial_x u$, as will

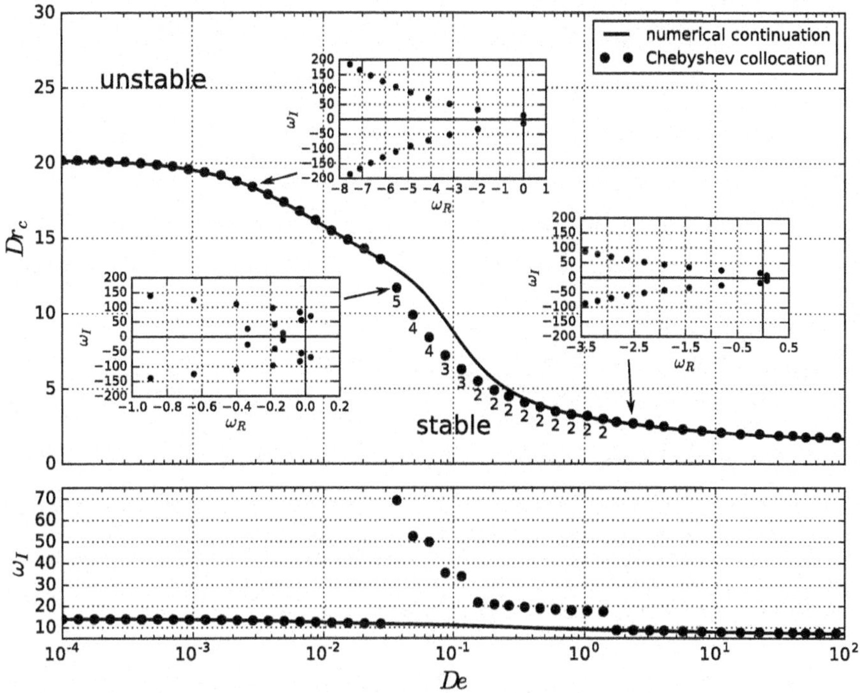

Fig. 8.9 Comparison of the PTT stability curves for $\varepsilon = 1$ as calculated by numerical continuation and by eigenvalue spectrum analysis via Chebyshev collocation. The number below the points indicates the number of the leading perturbation mode and is omitted for the first mode. The spectra at three different Deborah numbers are shown by insets

be discussed in Sect. 8.2. The dependence of the parameters p and q on the nonlinear parameter ε are shown in Fig. 8.12. The critical exponent p decreases with increasing ε, which can be fitted by a linear function as shown in Fig. 8.12a. Comparison with the results of the Giesekus model reveals that while the critical exponent is comparable for small values of the nonlinear parameters, it shows diverse trends for larger α and ε, as the Giesekus model predicts an increase of p with increasing α. Consequently, the stability curves of the Giesekus model decay more steeply if α is increased, while they decay less steeply if ε is increased in the case of the PTT model. Parameter q decreases linearly with increasing ε for $\varepsilon < 0.5$ and is constant for larger values, with a plateau close to 1.4. A linear fit for $\varepsilon < 0.5$ is shown in Fig. 8.12b.

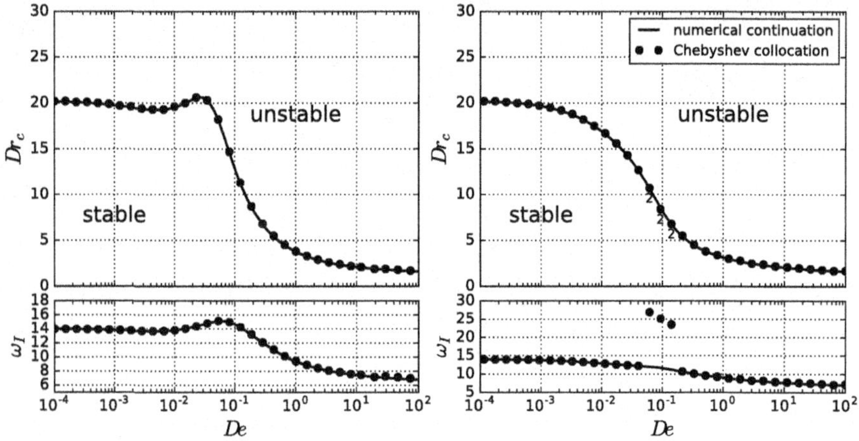

(a) $\varepsilon = 0.1$.

(b) $\varepsilon = 0.5$.

Fig. 8.10 Comparison of the PTT stability curves for $\varepsilon = 0.1$ and 0.5 as calculated by numerical continuation and by eigenvalue spectrum analysis via Chebyshev collocation. While for $\varepsilon = 0.1$ the first perturbation mode is the most unstable one for the entire range of De, the second mode partly dominates the stability behavior for $\varepsilon = 0.5$, as indicated by the number 2 below the dots. This leads to a noncontinuous trend of the oscillation frequency ω_I

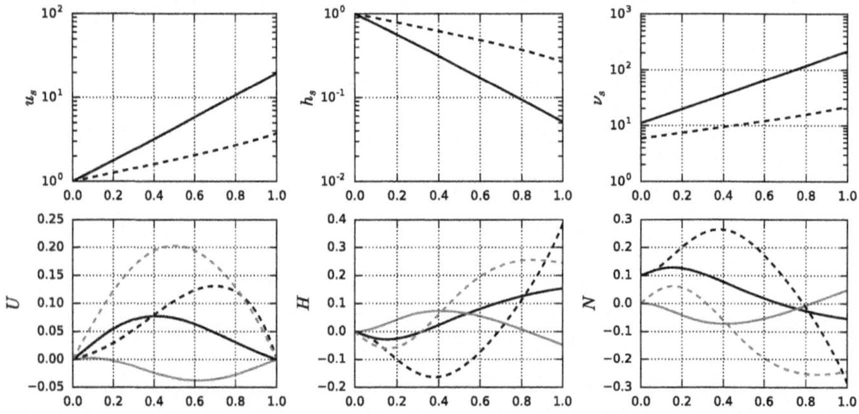

Fig. 8.11 Steady states and perturbations at criticality for the infinite-width PTT film casting model with $\varepsilon = 0.1$. The solid lines correspond to $De = 10^{-2}$ and the dashed lines to $De = 1$. The real parts of the complex perturbations are plotted in black and the imaginary parts in gray

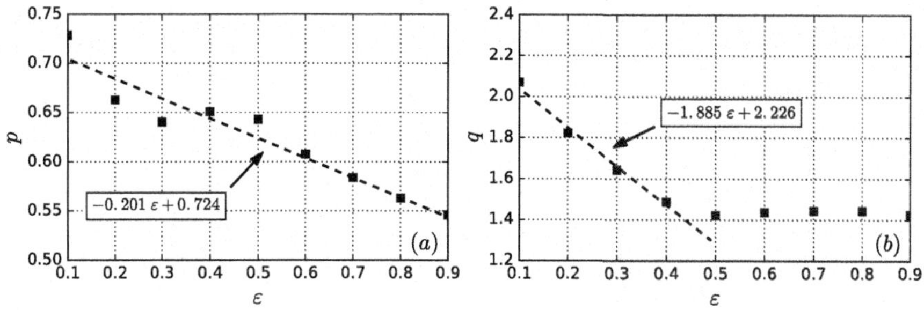

Fig. 8.12 Dependence of the fit parameters p and q of (8.8) on ε

8.1.4 Comparison and Evaluation of Constitutive Equations

For the remaining stability analyses of viscoelastic fluids, one of the previously studied constitutive equations needs to be chosen. Table 8.1 summarizes the main properties of the single stability predictions of the UCM, Giesekus and PTT models, so that a comprehensive evaluation becomes possible.

While the Giesekus and PTT models both yield similar qualitative results in that way that elasticity, i.e., increasing the Deborah number, has a mainly destabilizing effect, the UCM model leads to a fundamentally different picture. The predicted strong stabilization leads to both a second, higher critical draw ratio and to a complete suppression of draw resonance for large values of De. However, this behavior has

Table 8.1 Summary of the stability predictions of the various viscoelastic constitutive equations

Constitutive equation	Effect of elasticity (increasing De)	Behavior for $De > 0.1$	$Dr_{c,\infty}$	Remarks
UCM	Strongly stabilizing	No draw resonance	∞	Predicts second critical draw ratio, i.e., second stable regime
Giesekus	Mainly destabilizing, stability maximum for $\alpha \leq 0.1$	Dr_c exp. decaying	5.6	Incr. α mainly destabilizing, num. continuation correct only for $\alpha \leq 0.8$
PTT	Mainly destabilizing, stability maximum for $\varepsilon \leq 0.1$	Dr_c exp. decaying	1.7	Incr. ε purely destabilizing, num. continuation correct only for $\varepsilon < 0.5$

never been observed experimentally (see Sect. 1.3.2) and the available data exhibit clearly the trend of the other two viscoelastic models. Therefore, the UCM model can be identified as inappropriate for draw resonance studies. The main reason for this, as will become clear in Sect. 8.2, is the excessive strain hardening prediction also shown in Fig. 7.1a.

Comparing the behavior of the Giesekus and the PTT models reveals that the Giesekus model is very suitable for an analysis using numerical continuation, as the first perturbation mode is almost always the most unstable one, with the error in the critical draw ratio introduced for $\alpha = 0.9$ being small.

8.2 Effective Elongational Viscosity

Comparing the steady planar elongational viscosity predictions shown in Fig. 7.1 with the corresponding stability curves, i.e., Figs. 8.1, 8.3 and 8.8, may lead to the assumption that there is a link between the non-Newtonian viscous behavior of the material and the critical draw ratio. For instance, parameter combinations for which strain hardening is predicted often lead to a stability maximum. However, there are exceptions, e.g., the PTT model with $\varepsilon = 0.3$ predicts strain hardening, but increasing the Deborah number has a purely destabilizing effect.

One of the most extensive investigations on the connection between viscosity and stability was presented by Iyengar and Co (1993, 1996), who correlated both the shear and elongational viscosities to the stability curves, but their approach suffers from some drawbacks. While Iyengar and Co analyzed an infinite-width film casting process, they used the uniaxial elongational viscosity for the discussion of the mechanism. Instead, the planar elongational viscosity, which is in contrast to the Newtonian case not linked in a simple way to the uniaxial one, should be used. Even more important, however, is the fact that the viscosity does not depend on the deformation only, but is in general a time-dependent quantity. The often discussed steady trend reflects the material behavior after all relaxation processes are completed. In a process like film casting or fiber drawing, this is certainly not the case and the transient properties of the viscosity need to be accounted for.

For all calculations presented within this book, it is assumed that the material is completely relaxed at the inlet ($x = 0$).[3] Following a single fluid element traveling from the inlet to the outlet, it is possible to define a corresponding viscosity at every point on the x-axis, which depends on both the elapsed time and the present and past deformation. In other words, the time-dependence of the elongational viscosity is implicitly connected to the strain rate by the process itself, namely by the fluid velocity profile. As a consequence, it is possible to evaluate this *effective elongational*

[3] In practice, this point of largest swelling is located at some downstream distance from the die, as the material leaves the die with a certain deformation history due to the extrusion process.

viscosity μ^{eff}, also often referred to as apparent viscosity, by analyzing the steady state solutions[4] obtained via numerical continuation, and to examine the strain rate dependence of this viscosity.

As the focus of this book lies on the stability of the system, in particular on the critical parameters, the effective viscosity is evaluated for various values of De along the neutral stability curve, i.e., for $Dr = Dr_c$. For every value of De, and therefore for a particular value of Dr, a certain range of strain rates and corresponding effective viscosities is obtained. It turns out that those single parts of the viscosity curve can be combined to one smooth mastercurve, if μ^{eff} is plotted against $De\, u_s'$. Further tests revealed that even for draw ratios above or below criticality, the resulting effective viscosity curves lie on this mastercurve. Figure 8.13 shows these mastercurves for the UCM, PTT and Giesekus models.

In the case of the UCM model, the values of De belonging to the folding point and the upper branch of the stability curve are marked correspondingly. In contrast to the steady[5] viscosity curve, the effective viscosity shows a small minimum, similar to the small minimum of the neutral stability curve. Moreover, this analysis reveals that while the steady viscosity diverges at $De\, u_s' = 1/2$, the effective viscosity remains finite for all analyzed values of the stability curve. As a consequence, the divergence cannot be correlated with the critical Deborah number, beyond which draw resonance is completely suppressed.

The plots for the PTT and Giesekus models show small deviations of the single parts from the mastercurves, especially around $De = 1$. These deviations always belong to data points corresponding to x values close to the inlet. Modification of the initial stress boundary condition (7.20) clearly influences the sign and magnitude of the deviations, and it is strongly assumed that they are the result of the imperfect guess of the initial values of the stress components (see Sect. 7.3.2). The numerical values appearing in (7.20) have been chosen such that the deviations become minimal for the PTT model, as this model was chosen as the main model for this chapter. For this reason, the error is more pronounced in the case of the Giesekus model. Apart from that, it is quite striking that, in contrast to the steady viscosity curves, the effective viscosity shows a maximum if and only if the corresponding stability curve has a maximum as well.

All these observations strongly suggest that the effective elongational viscosity could be the key for understanding the main stability mechanism of viscoelastic fluids in draw resonance. This supposition will be investigated further in the following. Prior to this, it is worth mentioning that the fact that there exist effective viscosity mastercurves independent of the draw ratio is far from being trivial. In fact, no comprehensive explanation for this observation could be found yet and it will be left to future studies to find a satisfying answer to this question.

---

[4] In particular the normal stress difference ν divided by the strain rate $\partial_x u$ at every point x.

[5] The term "steady" may lead to some confusion here, as it is used in two different contexts. The "steady viscosity" denotes the viscosity curve after all relaxation processes are completed. The "steady state solutions" are the time-independent state quantities, which are used to compute the effective viscosity.

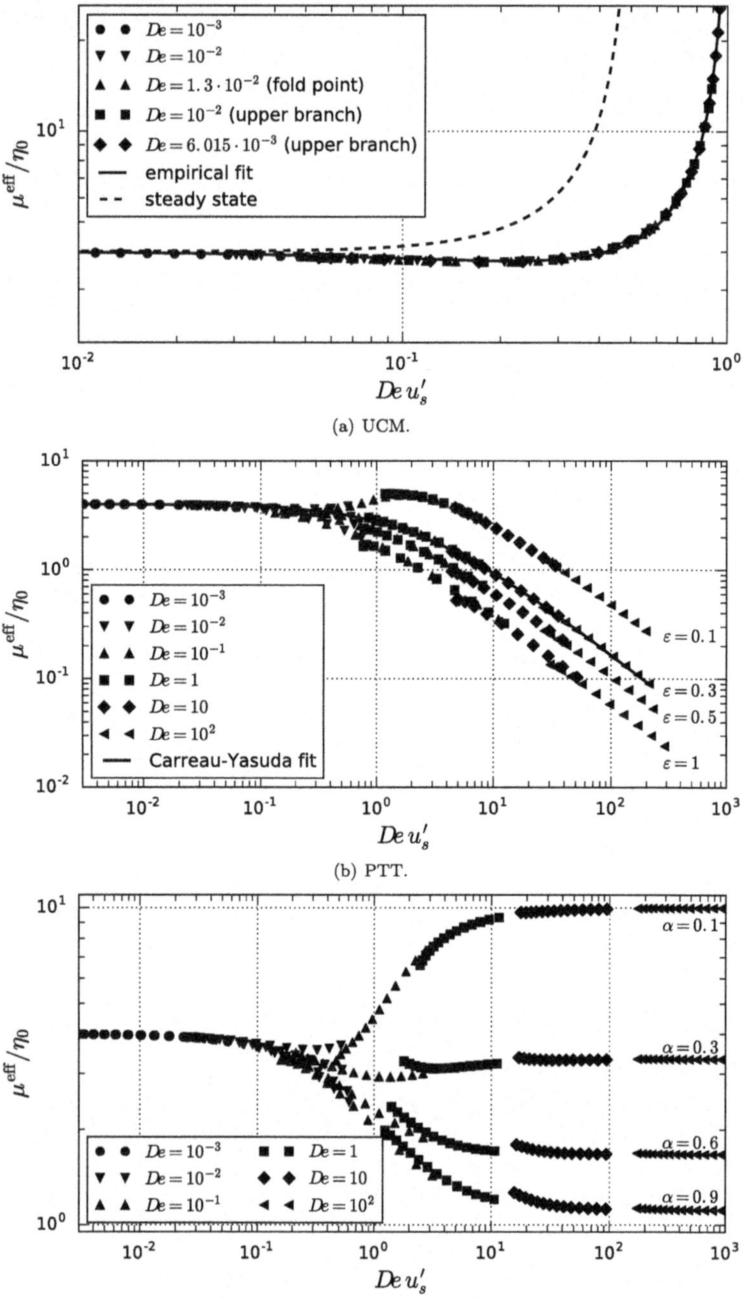

Fig. 8.13 a–c Effective elongational viscosity curves for the UCM, PTT and Giesekus models. In the case of the UCM model **a**, the steady planar elongational viscosity is shown as well, together with an empirical fit to the effective viscosity using (8.22). In the case of the PTT model **b**, the data belonging to $\varepsilon = 0.3$ is fitted with the Carreau–Yasuda function (8.18).

8.3 Generalized Newtonian Fluid Models

The previous findings on the effective elongational viscosity suggest the introduction of another group of non-Newtonian models, the so-called *generalized Newtonian fluid (GNF) models*. A GNF model uses the Newtonian constitutive equation (1.7), but with a viscosity depending on the deformation rate. This makes it possible to cover non-Newtonian effects caused by viscoelasticity without loosing the simple, explicit form of the constitutive equation. The drawback, however, is that these models are not capable of covering "true" elasticity,[6] as the corresponding correlation between stress and deformation is missing in the constitutive equation.

Instead of (1.7), GNF models directly involving the extensional process variables will be used here, i.e.,

$$\nu = \mu\left(\partial_x u\right)\partial_x u\ , \tag{8.12}$$

with a yet arbitrary elongational viscosity function $\mu\left(\partial_x u\right)$. The film model then consists of (8.12), continuity Eq. (2.33) and the purely viscous momentum equation,

$$\partial_x\left(h\nu\right) = 0\ . \tag{8.13}$$

The main idea is to check the assumption that (non-Newtonian) viscous effects are responsible for the viscoelastic stability behavior. If this is the case, a GNF model relying on the effective viscosity should be sufficient to reproduce the essential stability results. Two prominent examples of GNF models are introduced in the following.

8.3.1 The Power-Law Model

The most simple GNF model is the so-called *power-law model*.[7] It has the following form,

$$\mu_{\mathrm{PL}}\left(\partial_x u\right) = \mu_{\mathrm{PL},0}\left(\lambda_{\mathrm{PL}}\partial_x u\right)^{n-1}\ , \tag{8.14}$$

---

[6] The term "true elasticity" is used here to distinguish from non-Newtonian viscous effects, which are also induced by elasticity.

[7] The draw resonance behavior of the power-law model was already investigated very early (Pearson and Shah, 1974; Aird and Yeow, 1983) and this section is mostly a review of these results.

with power-law index n, a characteristic time λ_{PL} and a specific viscosity $\mu_{\mathrm{PL},0}$, which is equal to μ_{PL} if $\lambda_{\mathrm{PL}}\partial_x u = 1$. The characteristic time is primarily introduced to avoid problems with defining the dimensions of $\mu_{\mathrm{PL},0}$. Due to the form of (8.12), both parameters can be scaled out[8] in the momentum Eq. (8.13) and the only remaining control parameter is the power-law index n.

The steady state equations for the power-law model can then be written as

$$u_s' = \left(\frac{\nu_s}{4}\right)^{1/n} , \tag{8.15a}$$

$$h_s' = -\frac{h_s}{u_s}\left(\frac{\nu_s}{4}\right)^{1/n} , \tag{8.15b}$$

$$\nu_s' = \frac{1}{u_s}\left(\frac{\nu_s^{1+n}}{4}\right)^{1/n} , \tag{8.15c}$$

and the perturbation equations are given by

$$U' = \left(\frac{\nu_s}{4}\right)^{1/n}\frac{1}{u_s}\left(\frac{N}{n} - U\right) , \tag{8.16a}$$

$$H' = -\left(\omega\frac{H}{u_s} + U\right) , \tag{8.16b}$$

$$N' = -H' . \tag{8.16c}$$

This system of Eqs. (8.15) and (8.16) is solved by continuation with the boundary conditions (3.5), (3.11c) and (3.18). The resulting neutral stability curves are shown in Fig. 8.14a. The critical draw ratio increases monotonically with increasing power-law index and for $n = 1$, the Newtonian value is reobtained. The data presented by Pearson and Shah (1974) is plotted as well and shows perfect coincidence. As $n < 1$ corresponds to pure strain thinning and $n > 1$ to pure strain hardening, this simple model already gives an insight into the destabilizing or, respectively, stabilizing effect of these two scenarios. The dependence of the critical draw ratio on the power-law index can be approximated by an exponential function, as depicted in Fig. 8.14b. The PTT model predicts a power-law regime for large Deborah numbers (see Fig. 7.1c, with De scaling for $\lambda\partial_x u$).[9] As the power-law model is independent of the Deborah number, this could be used as an explanation for the occurrence of the plateau value $Dr_{c,\infty}$ in Fig. 8.3. However, the fact that the plateau value is predicted by the Giesekus model as well does not really fit in this picture, as the effective viscosity becomes constant there for large values of De. This issue will be discussed later in more detail.

[8] ν is then scaled by $\mu_{\mathrm{PL},0}\lambda_{\mathrm{PL}}^{n-1}\left(\frac{u_0}{L}\right)^n$.

[9] This is also true for the effective viscosity, as can be seen in Fig. 8.13b.

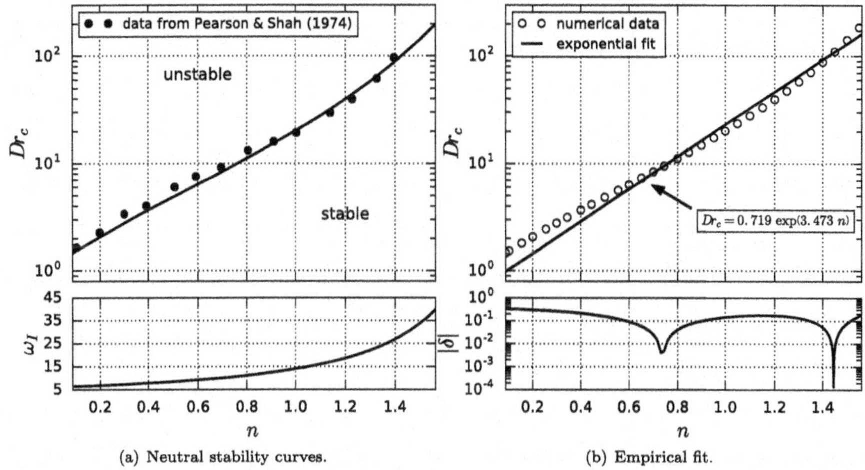

Fig. 8.14 Influence of the power-law index n on the critical draw ratio and the dimensionless oscillation frequency at criticality. **a** Neutral stability curves as calculated with numerical continuation (*lines*) and the results of Pearson and Shah (1974) (*points*). **b** Empirical fit for the critical draw ratio. Only few points are shown here for the sake of readability. The relative deviation to the numerical data $|\delta|$ is shown at the bottom and $R^2_{\text{adj}} = 0.946$

8.3.2 The Carreau–Yasuda Model

The biggest drawback of the power-law model is that it is only applicable for sufficiently large deformation rates. The viscosity diverges for small values and is not defined for the deformation-free state. The *Carreau–Yasuda (CY) model* (Macosko, 1994) provides a way to circumvent this obstacle. It can be written in the following form,

$$\mu_{\text{CY}}(\partial_x u) = \frac{4\,\eta_0}{\left(1 + (\lambda_{\text{CY}} \partial_x u)^m\right)^{\frac{n-1}{m}}} \,. \tag{8.17}$$

Similar to λ_{PL} in the power-law Eq. (8.14), λ_{CY} is a characteristic time. For $\partial_x u \ll \lambda_{\text{CY}}^{-1}$, the elongational viscosity reaches the planar Newtonian value $4\eta_0$, while for $\partial_x u \gg \lambda_{\text{CY}}^{-1}$, a power-law behavior with power-law index n is obtained. The region of transition can be tuned by the inner exponent m. In order to relate the CY model to the effective viscosities presented in Sect. 8.2, the characteristic time λ_{CY} will be split up into a product of a relaxation time and a characteristic parameter k, so that the rescaled form of (8.17) reads

$$\mu_{\text{CY}}(\partial_x u) = \frac{4}{\left(1 + (k\,De\,\partial_x u)^m\right)^{\frac{n-1}{m}}} \,, \tag{8.18}$$

where k indicates the reciprocal value of $De\,\partial_x u$, at which the transition from the Newtonian plateau to the power-law regime takes place. Note that this is an empirical ansatz for introducing a relaxation time and the Deborah number, which somewhat misuses their original meaning. Usually, the Deborah number can be seen as a measure for elasticity. A GNF model, however, does not cover any elastic effects, but only the non-Newtonian viscous effects which accompany the increasing influence of elasticity.

The stability behavior resulting from the CY model will be discussed later in this subsection and in Sect. 8.3.3. At this point merely the steady state and perturbation equations will be given. As the form of (8.18) makes it impossible to analytically solve (8.12) for $\partial_x u$, $\partial_x u$ will be used instead of v as state variable. The steady state is then governed by

$$u_s'' = \frac{u_s'^2\left(1 + k\,De\,u_s'\right)^m}{u_s c_{\mathrm{CY}}}\,, \tag{8.19a}$$

$$h_s' = -\frac{h_s}{u_s}u_s'\,, \tag{8.19b}$$

with

$$c_{\mathrm{CY}} = \left(1 + n\left(k\,De\,u_s'\right)\right)^m\,. \tag{8.20}$$

The perturbation equations read

$$\begin{aligned} U'' =& \frac{1}{u_s^2 c_{\mathrm{CY}}^2}\left[m(1-n)\left(k\,De\,u_s'\right)^m u_s'^2 U + \omega c_{\mathrm{CY}}\left(1 + \left(k\,De\,u_s'\right)^m\right) u_s' H\right. \\ &\left. + (1-n)\left(k\,De\,u_s'\right)^m (m + 2c_{\mathrm{CY}})\,u_s u_s' U'\right]\,, \end{aligned} \tag{8.21a}$$

$$H' = -\left(\omega\frac{H}{u_s} + U\right)\,. \tag{8.21b}$$

8.3.2.1 Comparison with UCM Model

In order to check whether it is possible to reproduce the stability results of the UCM model with a GNF model, the effective viscosity as shown in Fig. 8.13a is fitted with an empirical function of the following scaled form,

$$\mu_{\mathrm{UCM}}^{\mathrm{eff}}\left(De\,\partial_x u\right) = \left(\frac{2 + p_2}{1 - p_0 De\,\partial_x u} + \frac{2 - p_2}{1 + p_1 De\,\partial_x u}\right)\,, \tag{8.22}$$

which is inspired by (7.5) and has three free parameters, p_0, p_1 and p_2. The fitted curve is plotted in Fig. 8.13a as well, and the parameters were determined as $p_0 = 1.0$,

$p_1 = 1.64$ and $p_2 = -0.56$, with negligible errors and $R^2_{\text{adj}} = 1.0$.
Employing (8.22) in (3.1a), (8.12) and (8.13) leads to the following steady state equations,

$$u_s' = \frac{2\nu_s}{4 + \sqrt{16 + De\,\nu_s\,(p_0 + p_1)\,(4p_2 + De\,(p_0 + p_1)\,\nu_s)}}\,, \tag{8.23a}$$

$$h_s' = -\frac{h_s}{u_s} u_s'\,, \tag{8.23b}$$

$$\nu_s' = -\frac{\nu_s}{h_s} h_s'\,, \tag{8.23c}$$

and perturbation equations

$$U' = \frac{\nu_s\,(1 - p_0 De\,u_s')^2\,(1 + p_1 De\,u_s')^2}{4u_s\left[1 - De\,u_s'\left(\left(1 - \frac{p_2}{2}\right)\left(1 - De\,u_s' \frac{p_0}{2}\right) p_0 - \left(1 + \frac{p_2}{2}\right)\left(1 + De\,u_s' \frac{p_1}{2}\right) p_1\right)\right]} N \tag{8.24a}$$

$$- \frac{u_s'}{u_s} U\,,$$

$$H' = -\left(\omega \frac{H}{u_s} + U\right)\,, \tag{8.24b}$$

$$N' = -H'\,,$$

completed by the boundary conditions (3.5), (3.11c) and (3.28). Before looking at the neutral stability curves, the steady state predictions of the UCM model and its GNF approximation will be compared. Figure 8.15 shows the velocity and thickness profiles for both models for two values of the Deborah number, $De = 5 \cdot 10^{-3}$ and 10^{-2}, and for $Dr = 80$. The predictions coincide perfectly. More parameter com-

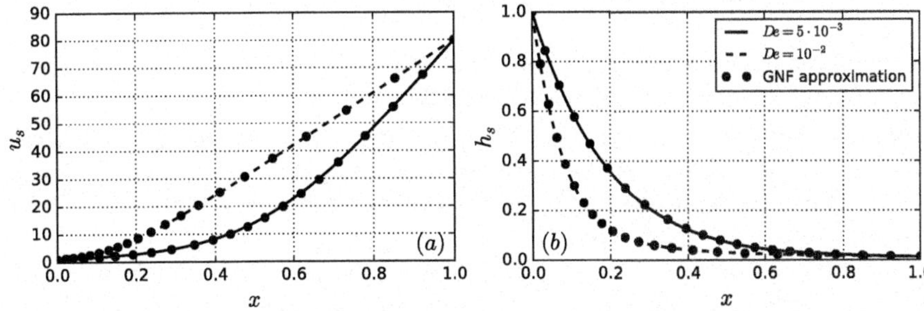

Fig. 8.15 Comparison of the steady state solutions of the UCM model and its GNF approximation for two values of the Deborah number and for $Dr = 80$. Both model predictions coincide perfectly

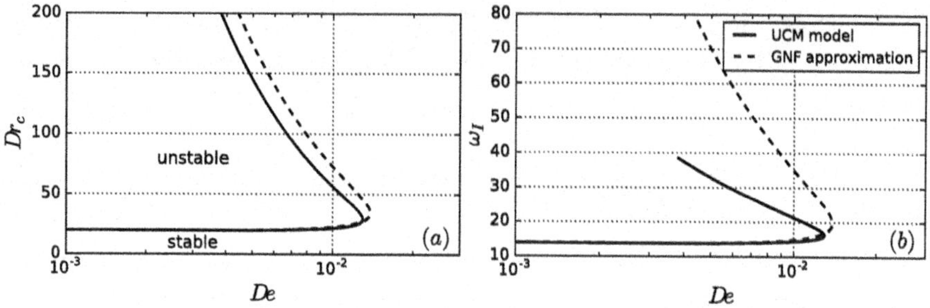

Fig. 8.16 Comparison of **a** the critical draw ratios and **b** the oscillation frequencies at criticality of the UCM model and its GNF approximation. The trends are comparable and the mismatch of the critical draw ratios is rather small

binations of *De* and for *Dr* have been checked as well such that it is possible to state that the steady state solutions can be reproduced by the GNF model without restrictions.

The stability predictions of the UCM model and the GNF approximation are compared in Fig. 8.16. The critical draw ratio shown in Fig. 8.16a resulting from the GNF model clearly has the same qualitative behavior than the UCM model, i.e., the strong stabilizing influence of the Deborah number and the nose-like shape of the curve, which leads to a second critical draw ratio. Moreover, the quantitative deviations are rather small. The oscillation frequencies at criticality plotted in Fig. 8.16b also show qualitative similarities, but the error of the GNF approximation is larger, especially for the upper branch. Nevertheless, both results for steady state and stability predictions suggest that the GNF approximation based on the effective elongational viscosity (8.22) is capable of reproducing the essential mechanism underlying the dynamics of the UCM model.

8.3.2.2 Comparison with PTT Model

In order to analyze the reproducibility of the PTT model results using a GNF approximation, the nonlinear parameter will be restricted to $\varepsilon = 0.3$ here. In this case, the effective viscosity as shown in Fig. 8.13b does not possess a maximum and can be fitted with the CY expression (8.18). The corresponding curve is plotted in Fig. 8.13b. The free parameters are determined as $n = 0.162 \pm 0.002$, $m = 0.889 \pm 0.003$ and $k = 0.433 \pm 0.003$, and $R^2_{\text{adj}} = 1.0$.

By utilizing these parameter values in (8.19), the steady state solutions were evaluated at several fixed values of the Deborah number and compared to the PTT model results. Figure 8.17 compares the velocity and thickness profiles for $Dr = 10$ at two different values of De.[10] Similar to the previous results for the UCM model, the GNF

[10] For both the UCM and PTT analyses, the value of Dr was chosen such that it lies in the range of critical draw ratios corresponding to the selected Deborah numbers.

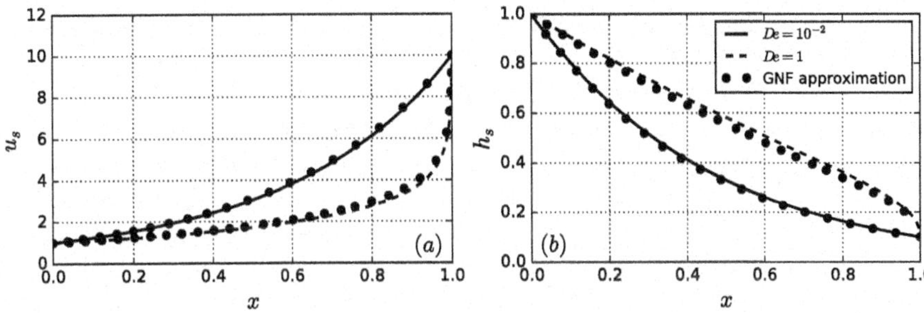

Fig. 8.17 Comparison of the steady state solutions of the PTT model and its GNF approximation for two values of the Deborah number and $Dr = 10$. Both model predictions coincide perfectly

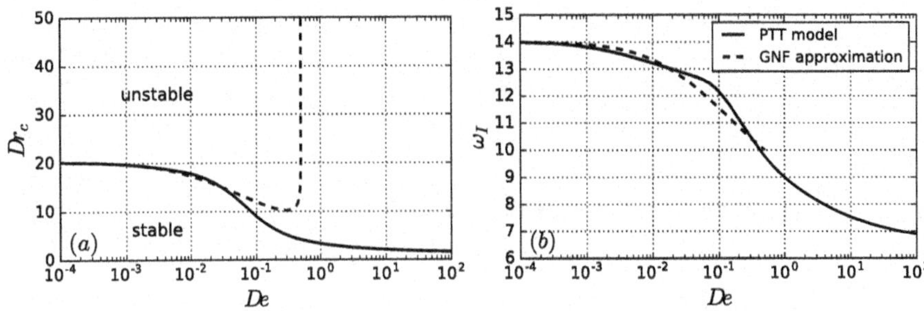

Fig. 8.18 Comparison of **a** the critical draw ratios and **b** the oscillation frequencies at criticality of the PTT model and its GNF approximation

model on basis of the CY viscosity function (8.18) shows very good agreement with the solutions of the PTT model. Small discrepancies can be explained by the small imperfections of the CY fit, which is itself only an empirical approximation to the effective elongational viscosity.

The situation changes significantly, however, for the stability results. Figure 8.18 depicts the neutral stability curves as calculated using the PTT model and its GNF approximation. While the critical draw ratios and the frequencies of both models are in quite good agreement for $De < 0.3$, the critical draw ratio as calculated using the GNF model starts to increase dramatically for larger Deborah numbers, and finally diverges at $De \approx 0.5$, which does not correlate with the PTT results in any way. On the other hand, the oscillation frequency does not vary significantly in the vicinity of $De = 0.5$ for both models.

In contrast to some scenarios presented in Sect. 8.1, the mismatch of the GNF approximation for $De > 0.3$ cannot be related to errors induced by numerical continuation, as higher perturbation modes were checked separately and do not become unstable at smaller draw ratios. Instead, there is a different explanation for this phenomenon, which will be given in detail in Sect. 8.3.3. In general, this result clearly indicates that the possibility to approximate the stability results of viscoelastic constitutive equations by a GNF model is limited at least to small Deborah numbers.

The reason for this lies most likely in the different form of the constitutive equations. While there exists a time derivative of the stress in the viscoelastic constitutive equations, the GNF equation yields merely an explicit expression for the stress. In the analysis of the steady state solutions, the time derivative is neglected and an approximation using a GNF model is possible even for high Deborah numbers. The stability analysis however shows that non-Newtonian viscous effects do not dominate the stability behavior over the whole range of Deborah numbers, i.e., the elastic part in the constitutive equation becomes important, which cannot be reproduced by the GNF equation. In the case of the UCM model, this issue did not appear as the entire stability curve is located at $De < 0.013$.

The supposition that "true" elastic effects can play a dominant role as well is supported by the observation that the effective viscosity of the Giesekus model (see Fig. 8.13c) has a plateau value for large values of $De\, u'_s$. If the critical draw ratio would be governed primarily by viscous effects, $Dr_{c,\infty}$ should be equal to the Newtonian value of 20.218, which is clearly not the case.

8.3.3 *Influence of Strain Hardening/Thinning*

The previously presented results show that the strain rate dependence of the elongational viscosity can be correlated to the stability behavior at least for small Deborah numbers. Motivated by the results of the power-law and UCM model analyses, it could be supposed that strain hardening always leads to an increase of the critical draw ratio, while strain thinning in turn leads to a decrease of the critical value. The results of the CY model (see Fig. 8.18a), however, show a dramatic increase of the critical draw ratio after a moderate decrease for increasing Deborah number, even though the CY model predicts a monotonically decreasing viscosity with increasing $De\, u'_s$, i.e., pure strain thinning, as visible in Fig. 8.13b for $\varepsilon = 0.3$. In the following, the mechanisms underlying the effects of strain hardening and thinning on draw resonance will be analyzed separately, in order to shed light on this particular issue.

The GNF model using the CY Eq. (8.18) was analyzed for several parameter combinations of n, m and k in order to find a reason for the unexpected behavior of the stability curve shown in Fig. 8.18a. It turns out that the stability curves do not always diverge, but sometimes merely reveal a stability maximum. Variation of the parameters indicates that the strength of this maximum, which may eventually end up in a divergence of the curve, can be tuned by the steepness of the power-law regime, i.e., by parameter n, and by the sharpness of the transition, i.e., by parameter m. The maximum grows hereby for decreasing n and increasing m. An example of a non-diverging stability curve is visualized in Fig. 8.19, where m and k have been set to unity for the sake of simplicity and $n = 0.3$. Increasing the Deborah number has mostly a destabilizing effect, and for large values of De, the power-law value of Dr_c for $n = 0.3$ is approached. A small stability maximum is located in the vicinity of $De = 1$. In contrast to the viscoelastic models discussed in Sect. 8.1, it must be possible to correlate the stability curve here with the non-Newtonian viscous

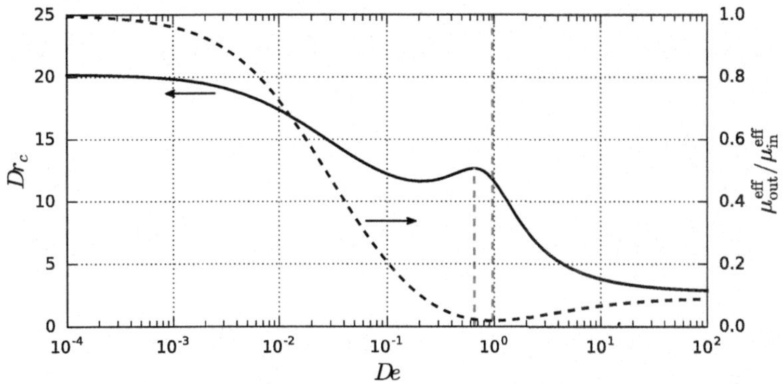

Fig. 8.19 Stability curve of the GNF model using the CY Eq. (8.18) for $n = 0.3$, $m = 1$ and $k = 1$ (*thick line*), together with the ratio of effective elongational viscosities at outlet to inlet (*dashed line*). The dashed gray lines indicate the maximum or, respectively, minimum of the two curves

behavior, as a GNF model is used, which is not capable of elastic effects. Recalling the stability mechanism of neck-in presented in Sect. 6.3, a correlation similar to (6.20) can be derived, but with the effective viscosity instead of the Trouton ratio. For this reason, the ratio of effective viscosities at outlet to inlet is plotted in Fig. 8.19 as well. Close to the stability maximum, this ratio becomes minimal, i.e., a maximum of damping of the feedback mechanism, which therefore can explain the stabilizing effect. Following this argumentation would lead to the statement that the stability increases if the effective viscosity at the outlet is smaller than that at the inlet. This, however, would mean that strain thinning should have a purely stabilizing effect, and strain hardening a destabilizing one, as the strain rate increases towards the outlet. As this is not the case, there must exist a second mechanism influencing the stability, which will be revealed in the following.

The argumentation above uses exclusively the steady state values of the effective viscosity. For this reason, the mechanism will be referred to as a *static mechanism*. The effective viscosity, however, depends on the strain rate, which is time-dependent itself. This means that a temporal variation of the strain rate also induces a temporal variation of the viscosity. Equation (3.31) can be generalized to

$$(\partial_x h)^{\mathrm{in}} = -\frac{f}{\left(\mu^{\mathrm{eff}}\right)^{\mathrm{in}}} \,. \tag{8.25}$$

The second, *dynamic mechanism* can be described now as follows. If the force increases temporarily,[11] the thickness close to the inlet will decrease due to (8.25). This will increase the velocity due to continuity, and therefore the strain rate increases as well. If the strain rate increases, the viscosity will decrease (increase) for a strain thinning (hardening) material. This in turn increases (decreases) the right-hand side

[11] The argumentation for a temporarily decreasing force is similar.

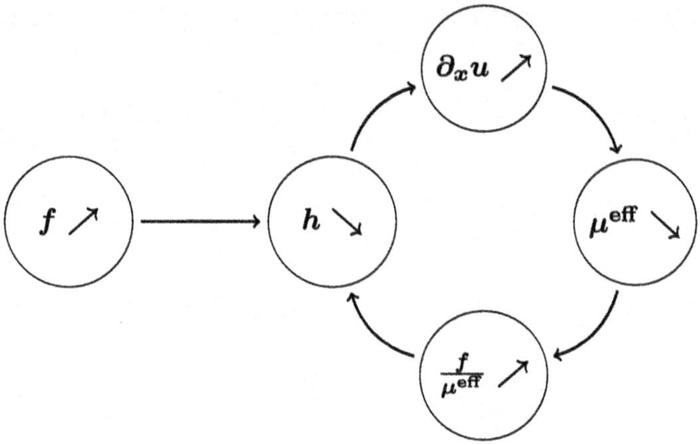

Fig. 8.20 Visualization of the dynamic destabilizing mechanism of strain thinning; h, $\partial_x u$ and μ^{eff} correspond to a position close to the inlet

of (8.25), which directly leads to an amplification (damping) of the initial force perturbation, and therefore to destabilization (stabilization). Figure 8.20 visualizes this mechanism in the case of strain thinning.

Whether the static or the dynamic mechanism governs the stability depends on the particular shape of the viscosity function. In the case of a power-law, the dynamic mechanism clearly dominates. In the case of the CY model, it is possible that the viscosity close to the inlet is governed by only weak non-Newtonian effects, while the viscosity close to the outlet already lies in the power-law regime, and is therefore a lot smaller, as the strain rate increases from inlet to outlet. Then the effect of the dynamic response of the viscosity can be too weak to compensate the effect caused by the ratio of steady viscosities at outlet to inlet. This also explains why the CY model leads to diverging stability curves if the transition to the power law is too sharp or the power-law index too small.

Employing a GNF model which uses a viscosity function depending only on the steady state of the strain rate yields indeed a purely stabilizing effect of strain thinning and a purely destabilizing effect of strain hardening. A similar effect exists in the non-isothermal case (see Chap. 10), where cooling, which usually stabilizes the process, becomes destabilizing for large heat transfer to the ambiant. In this regime, the temperature of the film is equal to the fixed and time-independent ambient temperature, which hinders the dynamic stability mechanism.

8.3.4 Elastic Effects and Stability Mechanism

The results presented in Sect. 8.3 clearly indicate that, besides (non-Newtonian) viscous effects, elastic effects may play a dominant role for the stability dynamics. The formulation of an explaining stability mechanism, however, is a non-trivial task. In contrast to all other mechanisms described throughout this book, it is not possible here to extend or modify the mechanism of a purely viscous Newtonian fluid (see Sect. 3.3), as the form of the constitutive equation changes fundamentally. The continuity equation cannot be used anymore to relate thickness perturbations at the inlet to a variation of the force, as there is no explicit form of the force available. It would be possible to use only the viscous force, i.e., $\mu h \partial_x u$, as cause for the inlet perturbations, but this force is not constant along the x-axis, as the momentum equation involves the total force, including the elastic part.

Another obstacle arises from the fact that with increasing elasticity, the time shift Δt_h between force and thickness perturbation waves becomes negative, which was already reported by Jung et al. (2000) and observed within the present investigations as well. However, a force which is delayed behind the thickness perturbations cannot cause these perturbations. As a consequence, the basic ansatz of the stability mechanism needs to be reconsidered and probably changed fundamentally. This extensive task will not be treated in the scope of this book and will be left to future studies. Nevertheless, it is worth mentioning that Kim's stability criterion (3.33) seems to be valid for viscoelastic constitutive equations as well, which has been reported by Jung et al. (2000) for a WM model, and which is exemplified in Fig. 8.21 for the PTT model, with $\varepsilon = 0.1$ and 0.3.

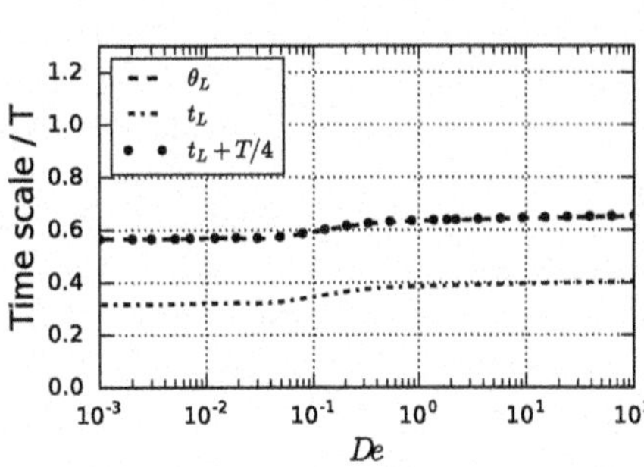

(a) $\varepsilon = 0.1$.

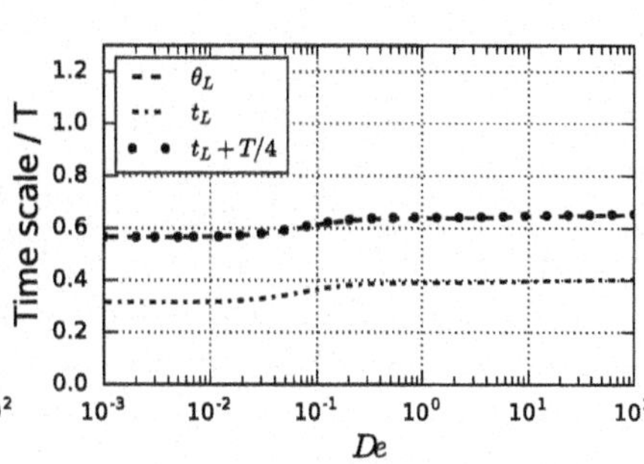

(b) $\varepsilon = 0.3$.

Fig. 8.21 Check of Kim's criterion for the PTT model at criticality. The time is scaled by the oscillation period T. Only few markers are shown for $t_L + T/4$ for the sake of readability

8.4 Conclusions on the Choice of the Constitutive Equation

The selection of an appropriate viscoelastic constitutive equation is always a challenging task. In order to shed light on the question about mandatory properties of this equation for a proper modeling of draw resonance, three infinite-width film casting models were analyzed by means of linear stability, namely the UCM model and two nonlinear extensions, the Giesekus model and the simplified PTT model. A careful model derivation was given in each case. Due to qualitatively wrong stability predictions, the UCM model is not recommended here. The other two models yield similar stability results, revealing in particular an exponential dependence of the critical draw ratio on the Deborah number for $De > 0.1$. The Giesekus model predicts unrealistic elongational viscosities for large Deborah numbers and/or strain rates, which could be a good reason to use the PTT model in many cases. A numerical continuation method, however, needs to be applied with care, as the most unstable mode may change for values of the nonlinear parameter $\varepsilon \geq 0.5$.

The effective elongational viscosity was also examined in the scope of draw resonance and mastercurves independent of the draw ratio were revealed. The neutral stability curves appear to have strong correlations with the effective elongational viscosity, which indicates the major importance of elongational properties predicted by the constitutive equation. This finding emphasizes the impracticality of the UCM model. GNF models on basis of the effective elongational viscosity can reproduce the steady state solutions of the viscoelastic models, but lead to wrong stability results for $De > 0.1$. The reason for this is most likely the inability of covering purely elastic effects, which appear to dominate the stability behavior for large Deborah numbers.

Finally, the mechanism underlying non-Newtonian viscous effects, i.e., strain hardening and thinning, turns out to be of higher complexity than one might think intuitively. Two effects, a static and a dynamic mechanism, act in opposite manner and the statement that strain hardening is always stabilizing and strain thinning always destabilizing is not generally true. An analysis of pure elastic effects is not straightforward due to the fundamentally different structure of the model equations and thus postponed. Nevertheless, Kim's stability criterion was successfully verified for the PTT model, providing further justification for its recommendation and its use in the following chapter.

Chapter 9
Practical Scaling and Stability Maps

Abstract While the effects of gravity, inertia, neck-in, surface tension and viscoelasticity were treated isolated in Chaps. 4–8, they mostly superimpose in practice. This chapter treats the application oriented analysis of models including up to four of these effects simultaneously. For this purpose an alternative scaling is used to introduce control parameters with strong correlation to typical practical quantities. Newtonian models for infinite and finite-width film casting including inertia and gravity effects are analyzed as well as a fiber drawing model, which covers additionally the influence of surface tension. Moreover, the infinite-width film model and the fiber model are extended to cover viscoelastic effects by employing the PTT constitutive equation. For each model, stability maps in the control parameter space are presented and discussed, and empirical formulas, which enable a simple and quick estimation of the critical draw ratio in dependence of the practical control parameters, are derived for particular scenarios.

9.1 Practical Control Parameters

The commonly used control parameters Re, Re/Fr, Ca^* and De as defined by (2.6) and (7.9) may be of moderate use if it is desired to have a direct connection to experimentally variable parameters, e.g., the zero-shear viscosity η_0 or the inlet velocity u_0, where the latter for instance influences all four control parameters. As a consequence, a stability analysis using these control parameters will not yield correlations directly applicable in practice. Within this chapter, it is assumed that the length L is fixed for a particular setup, while parameters like the inlet velocity, the viscosity, the relaxation time or the die geometry are, besides the draw ratio, the primarily changeable quantities, being tuned by the material itself or by the extrusion settings. Instead of u_0, the velocity u is scaled by $\sqrt{gL}$, which is a constant velocity as the length is constant itself. In particular, this velocity corresponds to the speed of a fluid particle free-falling along the length L under gravity. The scaling rules (2.5) and (2.28) for the time t and the stress tensor $\boldsymbol{\sigma}$ are modified accordingly. For the Newtonian models, this alternative scaling merely implies a change of the momentum equation and the boundary conditions. For the viscoelastic models, the

© Springer Nature Switzerland AG 2025
M. Bechert and B. Scheid, *Draw Resonance*, Fluid Mechanics and Its Applications 141,
https://doi.org/10.1007/978-3-031-85568-9_9

Deborah number De is substituted additionally by the *dimensionless relaxation time* Λ, defined by

$$\Lambda = \sqrt{\frac{g}{L}}\lambda \,, \tag{9.1}$$

in all constitutive Eqs. (7.14), (7.17) and (7.19). The particular momentum equations change as follows.

For the infinite-width film casting model, (2.35) transforms to

$$Fh\left(\partial_t u + u\partial_x u - 1\right) = \partial_x\left(h\nu\right) \,, \tag{9.2}$$

where the *dimensionless fluidity* F, defined by

$$F = \frac{\rho\sqrt{gL^3}}{\eta_0} \,, \tag{9.3}$$

is introduced as a new control parameter. This parameter tunes both inertia and gravity effects simultaneously. For $F = 0$, the purely viscous, or purely viscoelastic, model is recovered.

The aspect ratio A, which is introduced with the finite-width film model in Sect. 6.2, is unaffected by the modification of the scaling rules, and A can also be seen as a dimensionless half-die width. For finite-width film casting model, the momentum Eq. (6.10) becomes

$$Fh\ell\left(\partial_t u + u\partial_x u - 1\right) = \partial_x\left(h\ell\nu\right) \,. \tag{9.4}$$

In the case of fiber drawing, (2.24) is rescaled to

$$Fa\left(\partial_t u + u\partial_x u - 1 - S\frac{\partial_x a}{a^{3/2}}\right) = \partial_x\left(a\nu\right) \,, \tag{9.5}$$

where the *dimensionless surface tension parameter* S, defined by

$$S = \left(\frac{\ell_c}{\sqrt{2R_0L}}\right)^2 \,, \tag{9.6}$$

is used, which compares the capillary length $\ell_c = \sqrt{\frac{\gamma}{\rho g}}$ with the geometric mean of the fiber dimensions. While the value of the fluidity regulates the effects of inertia, gravity and surface tension, the impact of the latter is also tuned by S, i.e., increasing surface tension of the material and decreasing die diameter enforces the influence of surface tension on the fiber dynamics.

For all models, the boundary conditions for the velocity appearing in (2.25), (2.36) or (6.12) are modified to

$$u(x = 0, t) = Q \,, \tag{9.7a}$$

$$u(x = 1, t) = Q\,Dr \,, \tag{9.7b}$$

with the *dimensionless inlet velocity* Q, defined by

$$Q = \frac{u_0}{\sqrt{gL}} \,. \tag{9.8}$$

In the viscoelastic case, the stress boundary condition at the inlet (7.20) is modified to

$$\left.\frac{\tau_{xx}}{\nu}\right|_{x=0} = 1 - \frac{1}{2}\exp\left(-2(Q\Lambda)^{0.7}\right) \,, \tag{9.9}$$

in the case of infinite-width film casting, and to

$$\left.\frac{\tau_{xx}}{\nu}\right|_{x=0} = 1 - \frac{1}{3}\exp\left(-2(Q\Lambda)^{0.7}\right) \,, \tag{9.10}$$

in the case of fiber drawing. The dimensionless inlet velocity appears in the boundary conditions only. If Λ and F are equal to zero, Q can have any value except zero and is fixed to unity throughout this chapter for the sake of convenience. The transformations to the former control parameters (2.6) and (7.9) are:

$$\Lambda Q = De \,, \quad FQ = Re \,, \quad \frac{F}{Q} = \frac{Re}{Fr} \,, \quad \frac{2FS}{Q} = \frac{1}{Ca^*} \,. \tag{9.11}$$

The reciprocal value of the surface tension parameter is in particular proportional to the rescaled Bond number $Bo^* = \alpha \frac{\rho g L^2}{\gamma}$, as $Bo^* = 1/(2S)$. In order to develop a feeling for the control parameters F, Q, S, Λ and A, Table 9.1 gives an overview of typical ranges for fiber spinning and film casting processes of polymer melts. To keep the estimation procedure as transparent as possible, typical ranges of the single process and material parameters are listed as well, which finally lead to the given boundaries of the dimensionless control parameters. It is distinguished between two types of fiber spinning processes. The first, denoted by "fiber spinning I", represents a common industrial process of rather large scale, while "fiber spinning II" stands for a small-scale setup, which is often used in laboratories. The experimental setup used for observing draw resonance in Sect. 1.4 is an example of this second type.

The parameter regions are visualized in the Q–F plane in Fig. 9.1 for all three processes. Especially the regions for fiber spinning I on the one hand, and fiber spinning II and film casting on the other hand form distinguished clusters, which already indicates that, despite the fact that the theoretical description of both fiber drawing and film casting are similar, the processes correspond to different process

Table 9.1 Realistic ranges of process and material parameters and resulting estimated ranges of the dimensionless parameters Q, F, S, Λ and A (in bold) for typical fiber spinning and film casting processes of polymer melts. "Fiber spinning I" corresponds to an industrial, productive process type, while "fiber spinning II" represents a small-scale, laboratory setup. The length L is assumed to be constant for a particular process, but can vary for different setups

Parameter	Definition	Unit	Fiber spinning I	Fiber spinning II	Film casting
L	–	m	0.5 – 3	0.01 – 0.3	0.01 – 1
η_0/ρ	–	m^2s^{-1}	0.1 – 1	1 – 10	1 – 10
γ/ρ	–	cm^3s^{-2}	10 – 40	10 – 40	–
λ	–	s	0.01 – 1	0.1 – 50	0.1 – 50
u_0	–	$m\,s^{-1}$	0.01 – 1	0.001 – 0.01	0.01 – 0.1
R_0	–	mm	0.1 – 0.5	0.5 – 2	–
ℓ_0	–	m	–	–	0.1 – 1
Q	$\frac{u_0}{\sqrt{gL}}$	**–**	$\mathbf{10^{-3} - 1}$	$\mathbf{10^{-4} - 10^{-1}}$	$\mathbf{10^{-3} - 10^{-1}}$
F	$\frac{\rho\sqrt{gL^3}}{\eta_0}$	**–**	$\mathbf{1 - 10^2}$	$\mathbf{10^{-4} - 1}$	$\mathbf{10^{-4} - 1}$
S	$\frac{\gamma}{2R_0\rho gL}$	**–**	$\mathbf{10^{-4} - 10^{-1}}$	$\mathbf{10^{-3} - 1}$	**–**
Λ	$\sqrt{\frac{g}{L}}\lambda$	**–**	$\mathbf{10^{-2} - 10}$	$\mathbf{10^{-1} - 10^3}$	$\mathbf{10^{-1} - 10^3}$
A	$\frac{\ell_0}{L}$	**–**	**–**	**–**	$\mathbf{10^{-1} - 10^2}$

parameter regions and, as it will be shown in the following sections, to different dynamical regimes. The overlap of "fiber spinning II" and "film casting" also shows that these two scenarios may reveal more similarities in their dynamical behavior than the two fiber spinning processes. Of course, further differences between fiber spinning and film casting, e.g., the influence of surface tension or neck-in, are not visible in this representation. Nevertheless, the Q–F plane appears to be the most suitable for visualizing the stability behavior in most cases in the following, as this enables a unified representation for both film casting and fiber spinning of Newtonian and viscoelastic fluids. Moreover, all four control parameters of (2.6) and (7.9) depend on Q and F according to (9.11).

It is impossible to summarize all film casting or fiber spinning processes, which could be described by the models presented here, by one parameter region, as the variety in setups is very large, including also processes of other materials, e.g., molten glass. For this reason, wherever possible a larger parameter space as compared to the depicted clusters in Fig. 9.1 is used for the process analyses in the following sections.

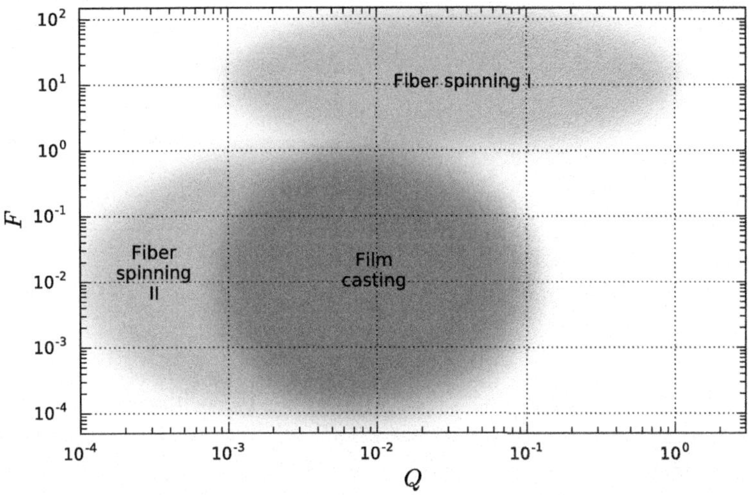

Fig. 9.1 Visualization of the parameter ranges of the fiber spinning and film casting processes described in Table 9.1 in the Q–F plane

9.2 Infinite-Width Film Casting

Within this section, a Newtonian and a viscoelastic infinite-width film casting model are analyzed with respect to fluidity, inlet velocity and relaxation time. The rather simple stability behavior of the Newtonian model makes it possible to derive empirical stability criteria for particular dynamical regimes. Stability maps in the control parameter space reveal those regimes for both models.

9.2.1 Newtonian Model

The results of the Newtonian infinite-width film casting presented here can also be used for fiber spinning processes with negligible surface tension. In this case the fluidity needs to be rescaled by 3/4 due to the different Trouton ratios, as explained in Chap. 2. Instead of the normal stress difference ν, the derivatives of the steady state velocity, u_s', and of the velocity perturbation, U', will be used here as state variables, as it turns out that this leads to a higher numerical robustness. This means that the normal stress difference (2.34) is directly substituted in the momentum equation (9.2). Mathematically, both approaches are equivalent. The resulting steady state equations are given by

$$h_s' = -\frac{h_s u_s'}{u_s} \,, \tag{9.12a}$$

$$u_s'' = -\frac{h_s' u_s'}{h_s} + \frac{F}{4}\left(u_s u_s' - 1\right) \,, \tag{9.12b}$$

and the perturbation equations read

$$H' = -\left(\frac{\omega}{u_s}H + U'\right) \,, \tag{9.13a}$$

$$U'' = \frac{\omega u_s'}{u_s^2}H + \frac{F}{4}\left[\left(\frac{1}{u_s} + \omega + u_s'\right)U + u_s U'\right] \,. \tag{9.13b}$$

A comparison of the calculated values using AUTO-07P with the values presented by Cao et al. (2005), who analyzed the same model, but with different definitions of the control parameters and for a smaller parameter range, is depicted in Table 9.2 and reveals almost perfect agreement. Figure 9.2 shows the dependence of the critical draw ratio on F for fixed values of Q (a), and on Q for fixed values of F (b). While increasing F has always a stabilizing effect, there exists a stability minimum with respect to Q, i.e., increasing Q can either have a destabilizing effect for small values

Table 9.2 Comparison of the calculated critical draw ratios Dr_c and frequencies ω_I using AUTO-07P with the values presented by Cao et al. (2005). Reproduced with permission from Bechert et al. (2015) and Elsevier

Process parameters		This book		Cao et al. (2005)	
F	Q	Dr_c	ω_I	Dr_c	ω_I
0	0	20.218	14.011	20.218	14.011
0.25	0.1	30.343	1.5895	30.343	1.5895
0.5	0.1	56.273	1.7732	56.267	1.7731

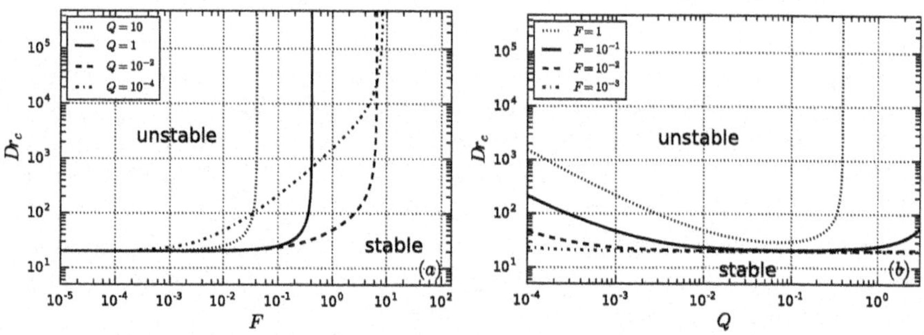

Fig. 9.2 Stability curves **a** Dr_c versus F for various values of Q and **b** Dr_c versus Q for various values of F for the infinite-width film casting model. Adapted with permission from Bechert et al. (2015) and Elsevier

of Q, or a stabilizing effect for large values. The purely stabilizing influence of the fluidity is not surprising, as both inertia and gravity effects are amplified with increasing F and both effects increase Dr_c, as it was shown in Chap. 4. Similar to the stability curve of the viscous–inertia model shown in Fig. 4.5a, all curves diverge and the system becomes unconditionally stable if F is above a threshold value, which is denoted by F_{us} and which increases with decreasing Q. Moreover, the stability curve corresponding to $Q = 10^{-4}$ shows that before the curve diverges, the critical draw ratio increases moderately with increasing F. The transition to unconditional stability can also be seen in the stability curves with respect to Q for $F = 1$.

In order to analyze the stability behavior further, the Dr_c versus F curves are fitted for various values of Q with the following empirical correlation function,

$$Dr_c = 20.218\left[1 + \left(\frac{BF}{F_{us} - F}\right)^C\right], \tag{9.14}$$

with free parameters B, F_{us} and C, which all depend on Q in general. Due to the diverging behavior of the data, the fitting was performed with Deming regression. As the errors in both variables appear to be only a few percent maximum, (9.14) has been found to be an appropriate choice.

The dependencies of the free parameters on Q are visualized in Fig. 9.3. While parameters B and F_{us} show a clear trend, C is close to unity for all values of Q with only minor variations. For this reason the focus of the following analysis lies on B and F_{us}. It is possible to fit the parameters with high accuracy using the following functions,

$$B(Q) = b_0\left[1 + \left(\frac{b_1}{Q}\right)^{b_2}\right], \tag{9.15a}$$

$$F_{us}(Q) = \frac{f_0}{Q/f_1 + 1}, \tag{9.15b}$$

with free parameters b_0, b_1, b_2, f_0 and f_1. The fits and the determined values of the free parameters are depicted in Fig. 9.3 as well. Both B and F_{us} show two distinct behaviors. For $Q \ll b_1$, B is almost inversely proportional to Q and for $Q \gg b_1$, B is constant and equal to b_0. Similarly, for $Q \ll f_1$, F_{us} is constant and equal to f_0, while it is inversely proportional to Q for $Q \gg f_1$. As b_1 and f_1 are very close, it is possible to distinguish between two general regimes, with transition from one to the other at $Q \approx 0.06$.

Consequently, for $Q \gg 0.06$, (9.14) can be approximated by

$$Dr_c \approx 20.218\left[1 + \left(\frac{b_0 F Q}{f_0 f_1 - F Q}\right)^C\right], \tag{9.16}$$

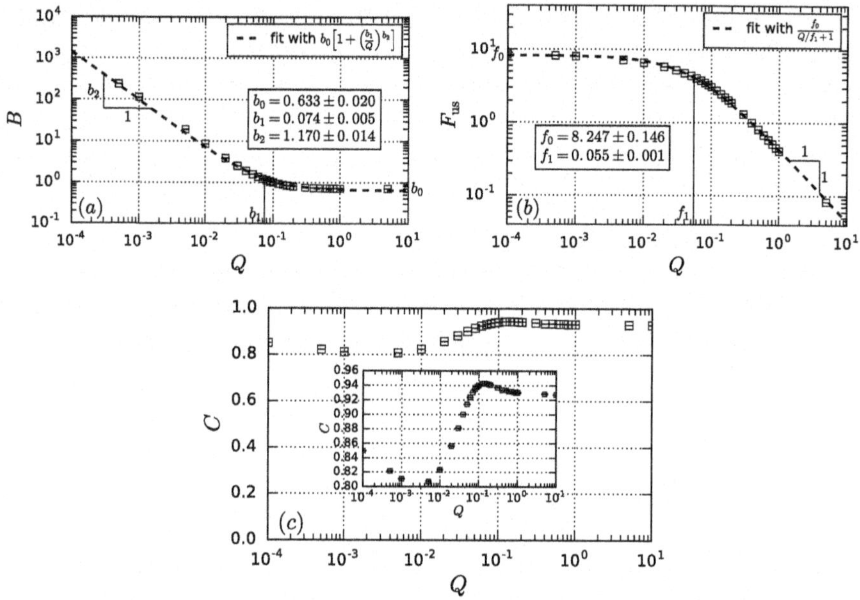

Fig. 9.3 **a–c** Dependencies of the fitting parameters of correlation (9.14) on the dimensionless inlet velocity Q, together with empirical fits. The slope triangles correspond to the logarithmic values and $R^2_{\mathrm{adj}} = 0.95$ for the fit in **a** and 1.0 for the fit in **b**. Adapted with permission from Bechert et al. (2015) and Elsevier

which indicates that the critical draw ratio depends on the product of F and Q only. According to (9.11), this product is equal to the Reynolds number and therefore it can be stated that the critical draw ratio is influenced exclusively by viscous and inertia effects. In fact, the coefficients of correlation (9.16) are almost identical to the ones obtained by fitting the viscous–inertia model with (4.8) (see Sect. 4.2).

If instead $Q \ll 0.06$, and if b_2 is further set to unity, (9.14) can be approximated by

$$Dr_c \approx 20.218 \left[1 + \left(\frac{b_0 b_1 F/Q}{f_0 - F} \right)^C \right]. \qquad (9.17)$$

If additionally $F \ll f_0$, the denominator can be substituted by f_0 and the critical draw ratio depends only on F/Q, which is according to (9.11) identical to Re/Fr. For $F < 0.8$, the error induced by neglecting F in the denominator is smaller than 10%. Within this particular region, the stability behavior is governed by viscous and gravity effects, and (9.17) coincides with (4.4), which was used to fit the stability curve of the viscous–gravity model (see Sect. 4.1).

A comprehensive overview of the influences of fluidity and inlet velocity on the critical draw ratio is given by the stability maps of Fig. 9.4. The left plot depicts isolines of constant Dr_c in the Q–F plane, with the particular value of Dr_c indicated below each line. The region of unconditional stability, where no draw resonance

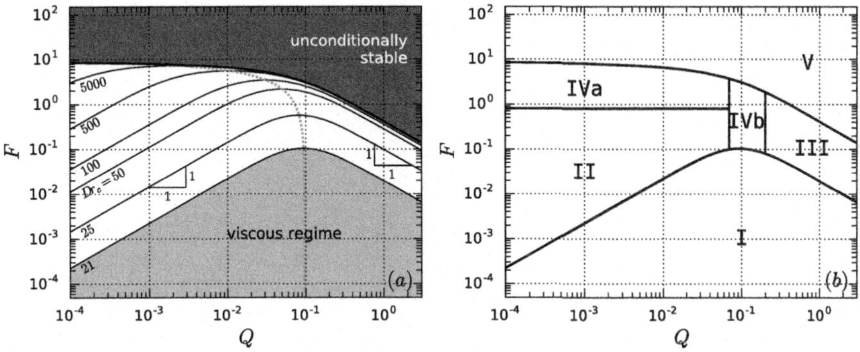

Fig. 9.4 Visualization of the stability behavior of the infinite-width film model with respect to dimensionless inlet velocity Q and dimensionless fluidity F. **a** Stability map for the infinite-width film model. The thin lines correspond to parameter combinations with constant critical draw ratio, $Dr_c = 21, 25, 50, 100, 500, 5000$. Within the light shaded region, Dr_c is between 20.218 and 21. The dark shaded region, which is bounded by a thick line, indicates unconditional stability, i.e., the system is stable for any value of Dr. The gray dotted line divides the parameter space into a region where increasing Q has a destabilizing effect (*left part*) and a region where it has a stabilizing effect (*right part*). The triangle slopes correspond to logarithmic values. **b** Partition of the parameter space into different regions. The corresponding regimes are explained in Table 9.3. Adapted with permission from Bechert et al. (2015) and Elsevier

occurs for any value of the draw ratio, is shaded in dark gray. Below the line for $Dr_c = 21$, the critical draw ratio is almost constant and approximately equal to the viscous limit of 20.218, and the corresponding region is shaded in light gray. The depicted 1-to-1 slopes for the logarithmic values of Q and F indicate that the critical draw ratio depends on either F/Q or FQ, i.e., a viscous–gravity or, respectively, a viscous–inertia regime. The gray dotted line indicates a stability minimum with respect to Q, i.e., increasing the inlet velocity has a destabilizing effect to the left of the line, while it has a stabilizing effect to the right of the line.

The right plot in Fig. 9.4 shows together with Table 9.3 the partition of the parameter space into the different dynamical regimes. Table 9.3 also gives empirical, analytical expressions for the critical draw ratio in dependence on F and Q. The threshold values for Q to the viscous–gravity–inertia region "IVb" correspond to the values of Q where the deviations from the reciprocal or, respectively, constant behavior of parameters B and F_{us} in Fig. 9.3 become larger than 10%. Gravity can always be neglected if $Q \gg 0.06$, as inertia then leads to unconditional stability for $F/Q = \mathcal{O}(1)$, where gravity still has a negligible influence on Dr_c. On the other hand, inertia can only be neglected for small Q if F is not too large as well. Otherwise, the stabilizing effect of inertia gets stronger and finally leads to an infinite critical draw ratio. In this case, however, the threshold value of the Reynolds number $Re = FQ$ decreases. This can be explained by the stability mechanism introduced in Sect. 4.2. As gravity already increases Dr_c, the threshold value for the effective draw ratio becomes larger.

Table 9.3 Empirical stability criteria for the various regimes shown in Fig. 9.4. $B(Q)$ and $F_{\mathrm{us}}(Q)$ are given in (9.15) and $C(Q)$ can be inferred from Fig. 9.3c. Adapted with permission from Bechert et al. (2015) and Elsevier

Label	Regime	$Dr_{\mathrm{c}}/20.218$
I	Viscous	1
II	Viscous–gravity	$1+\left(0.01\frac{F}{Q}\right)^{0.955}$
III	Viscous–inertia	$1+\left(\frac{0.703FQ}{0.414-FQ}\right)^{0.924}$
IVa	Viscous–gravity–inertia	$1+\left(\frac{0.0468F/Q}{8.247-F}\right)^{C(Q)}$
IVb	Viscous–gravity–inertia	$1+\left(\frac{B(Q)F}{F_{\mathrm{us}}(Q)-F}\right)^{0.94}$
V	Unconditionally stable	–

Comparing Figs. 9.4 and 9.1 reveals that the stability of most infinite-width film casting processes is governed by viscous and gravity effects, while it depends on the particular setup of fiber spinning (with negligible surface tension), whether the process is mostly unconditionally stable or not. In general there is only a small part of the parameter space, where both gravity and inertia need to be taken into account simultaneously.

9.2.2 *Viscoelastic Model*

The rescaled PTT constitutive Eq. (7.19) can be written as

$$\Lambda\left(\partial_t\overline{\nu}+u\partial_x\overline{\nu}+2\partial_x u\left(\overline{\nu}-2\overline{\tau}_{xx}\right)\right)+\mathrm{e}^{\varepsilon\Lambda(2\overline{\tau}_{xx}-\overline{\nu})}\overline{\nu}=4\partial_x u\ , \tag{9.18a}$$

$$\Lambda\left(\partial_t\overline{\tau}_{xx}+u\partial_x\overline{\tau}_{xx}-2\partial_x u\overline{\tau}_{xx}\right)+\mathrm{e}^{\varepsilon\Lambda(2\overline{\tau}_{xx}-\overline{\nu})}\overline{\tau}_{xx}=2\partial_x u\ , \tag{9.18b}$$

and will now be used to extend the infinite-width film casting model to cover viscoelastic effects. Besides the relaxation time Λ, the nonlinear parameter ε is added to the model as a consequence. In order to keep the analysis as simple as possible and to focus on the influence of the primary physical parameter, i.e., the relaxation time, ε will be fixed to 0.1 for all calculations. This choice is also motivated by the fact that no leading mode switching was observed in Sect. 8.1.3 for small values of ε.[1]

[1] Strictly speaking, the fact that no leading mode switching was observed for the viscous model without inertia and gravity does not automatically imply that this effect will be absent for the model used here, as gravity and inertia may influence the viscoelastic behavior, too. Nevertheless, it is a reasonable assumption and the second perturbation mode was tracked occasionally for some parameter combinations for cross-checking.

In contrast to the Newtonian model, it is not possible to relate the results of the viscoelastic film casting model directly to a fiber drawing model without surface tension effects, as the PTT constitutive equations differ non-trivially for planar and uniaxial elongations. A viscoelastic fiber drawing model including surface tension will be analyzed separately in Sect. 9.4.2. As there is no explicit expression for the stress tensor available, a direct substitution into the momentum equation, as it was done for the Newtonian model, is not possible here. Using (2.33), (9.2) and (9.18), the following steady state equations can be derived,

$$u_s' = \frac{1}{4\tilde{c} - F\Lambda u_s^2}\left(\frac{v_s}{\tilde{c}_{\mathrm{PTT}}} - F\Lambda u_s\right), \tag{9.19a}$$

$$h_s' = -\frac{h_s u_s'}{u_s}, \tag{9.19b}$$

$$v_s' = -v_s\frac{h_s'}{h_s} + F(u_s u_s' - 1), \tag{9.19c}$$

$$t_s' = \frac{1}{\Lambda u_s}\left(2u_s'(1 + \Lambda t_s) - \frac{t_s}{\tilde{c}_{\mathrm{PTT}}}\right), \tag{9.19d}$$

with

$$\tilde{c} = 1 + \Lambda\left(t_s - \frac{3}{4}v_s\right), \tag{9.20a}$$

$$\tilde{c}_{\mathrm{PTT}} = \mathrm{e}^{\varepsilon\Lambda(v_s - 2t_s)}. \tag{9.20b}$$

It turns out that the numerical robustness can be increased by using the unscaled stress perturbations, i.e., using the following perturbation ansatz for v and τ_{xx},

$$v(x,t) = v_s(x) + \tilde{N}(x)\,\mathrm{e}^{\omega t}, \tag{9.21a}$$

$$\tau_{xx}(x,t) = t_s(x) + \tilde{T}(x)\,\mathrm{e}^{\omega t}, \tag{9.21b}$$

with unscaled perturbations $\tilde{N} = v_s N$ and $\tilde{T} = t_s T$. It is assumed that the possible high gradients of the steady states of the stress components are the main reason why the alternative ansatz leads to better numerical manageability. If the former perturbation ansatz (3.24c) and (8.1) are used, the high gradients of the steady states lead to even larger gradients of the stress perturbations. For thickness and velocity, the usual ansatz (3.24a–3.24b) is employed. The perturbation equations can then be written as

$$
\begin{aligned}
U' = & \frac{1}{u_s\left(4\tilde{c} - F\Lambda u_s^2\right)} \Big[-\left(4\tilde{c}u_s' + \Lambda\left(v_s u_s' - u_s v_s' - F u_s^2(\omega + 2u_s')\right)\right) U \\
& + \Lambda\left(v_s(u_s' + \omega) - u_s v_s' + F u_s(u_s u_s' - 1)\right) H \qquad (9.22a) \\
& + \left(\frac{1 - \varepsilon\Lambda v_s}{\tilde{c}_{\mathrm{PTT}}} + \Lambda(\omega + 3u_s')\right)\tilde{N} + \left(\frac{2\varepsilon\Lambda v_s}{\tilde{c}_{\mathrm{PTT}}} - 4\Lambda u_s'\right)\tilde{T}\Big] ,
\end{aligned}
$$

$$
H' = -\left(\frac{\omega}{u_s}H + U'\right) , \qquad (9.22b)
$$

$$
\tilde{N}' = -v_s H' - \frac{h_s'}{h_s}\tilde{N} + F\left(u_s^2 U' + u_s(\omega + 2u_s')U\right) , \qquad (9.22c)
$$

$$
\begin{aligned}
\tilde{T}' = & \frac{1}{u_s\Lambda}\Big[2u_s(1 + \Lambda t_s)U' + \frac{t_s}{\tilde{c}_{\mathrm{PTT}}}U + \frac{\varepsilon\Lambda t_s}{\tilde{c}_{\mathrm{PTT}}}\tilde{N} \\
& - \left(\frac{1 + 2\varepsilon\Lambda t_s}{\tilde{c}_{\mathrm{PTT}}} + \Lambda(\omega - 2u_s')\right)\tilde{T}\Big] . \qquad (9.22d)
\end{aligned}
$$

Neutral stability curves with respect to either F, Q or Λ, where the remaining two other control parameters are kept fix, turn out to be of minor significance. Therefore it will be focused on the stability maps of isolines of the critical draw ratio in the Q–F plane, which are shown in Fig. 9.5 for four different values of Λ. For rather large values of the fluidity, no solutions could be calculated and the corresponding inaccessible regions are marked by big dots in the maps. Further analysis suggests that the calculation fails due to too large gradients of the steady states close to the outlet, which are assumed to be caused by an interplay of elastic and inertia effects. Based on the results of the Newtonian model, it would be reasonable to expect a region of unconditional stability due to inertia instead. While this may still be true in a pure mathematical sense, the fact that the numerical calculation breaks due to too large gradients strongly suggests that the process is not limited by draw resonance in this case, but by material failures, e.g., film break-up. While it is surely worth studying the impact of such events, it will be focused here on draw resonance only. As a consequence, the parameter space of the stability maps is diminished to $F \leq 1$, which still corresponds to the entire domain relevant for film casting according to Table 9.1. For $\Lambda = 100$, the limiting effect turns out to be more pronounced,[2] and the analysis is restricted to $F \leq 0.2$.

Apart from the inaccessible region, the stability maps also reveal other dynamical regimes. For small values of Q and Λ, and large values of F, the 1-to-1 slope of the isolines indicates that the critical draw ratio depends only on F/Q in this parameter region, and therefore this region is governed solely by viscous[3] and gravity effects. This means that within this particular region, the results of the Newtonian viscous–gravity model are fully applicable. With increasing relaxation time, a region

[2] In addition, viscosity and relaxation time are usually correlated, which means that the fluidity will decrease with increasing relaxation time in practice.

[3] If viscoelastic effects also contributed, changing Q would influence the isolines in a different way.

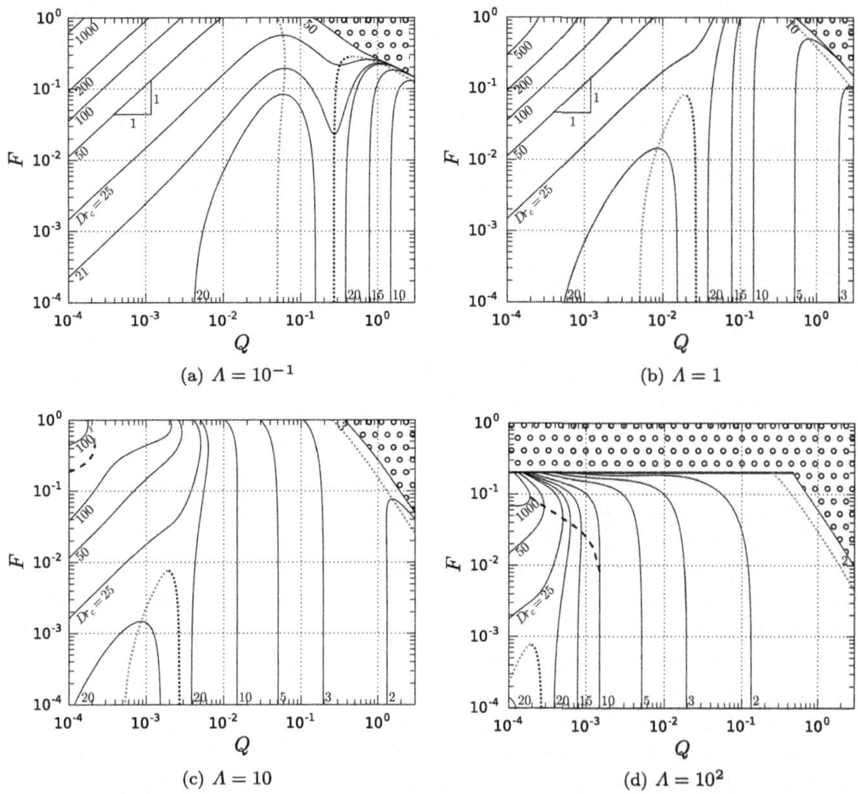

Fig. 9.5 a–d Stability maps for dimensionless relaxation times $\Lambda = 10^{-1}$, 1, 10, 10^2 for the PTT infinite-width film casting model. The thin lines correspond to parameter combinations with constant critical draw ratio, with particular values shown below or to the right of the lines. Within the dotted region, no solutions could be calculated, probably due to high gradients in the steady states. The black (resp. gray) dotted lines indicate maxima (resp. minima) of Dr_c with respect to Q for constant values of F. The black (resp. gray) dashed lines indicate stability maxima (resp. minima) with respect to F for constant values of Q

of vertical isolines grows from large values of Q towards smaller values. As this means that the critical draw ratio is independent of the fluidity, neither inertia nor gravity have an impact in this case, and the stability is determined by the viscoelastic model as presented in Sect. 8.1.3. In general, a large part of the relevant parameter space corresponds either to the viscous–gravity regime, or to the purely viscoelastic regime. For small F and Q values, gravity and viscoelastic effects may both be important if Λ is large, but inertia can still be excluded in this case as $Re = FQ$ is very small. In fact, inertia effects seem to play hardly a role, apart from the inaccessible region, and can be neglected in most cases. The direct visualization of the various dynamical regimes by the isolines of the critical draw ratio demonstrates the advantage of the practical mappings.

The maps in Fig. 9.5 show non-monotonic stability behavior for both Q and F. Extrema of the critical draw ratio with respect to Q for constant F and to F for constant Q were tracked with AUTO-07P and are depicted by dotted and, respectively, dashed lines. Maxima are drawn black, while minima are gray. The minimum and maximum with respect to Q, which can be tracked up to $F = 10^{-4}$, are caused by viscoelastic effects. For $F = 10^{-4}$, increasing Q basically follows the stability curve for $\varepsilon = 0.1$ in Fig. 8.8. For $\Lambda = 10$ and 100, a maximum with respect to F appears if Q is small. Both inertia and gravity, which are tuned by F, are known to have purely stabilizing effects. A possible explanation for this opposing effect is the coupling between viscoelastic effects and the stress components. With increasing fluidity, it is possible that the effects of gravity and inertia influence the stress components strongly enough so that viscoelastic effects are stimulated and lead to a decrease of the critical draw ratio.

9.3 Finite Width Film Casting

Besides fluidity F and inlet velocity Q, the aspect ratio A becomes important as additional control parameter when the finite width of the film is respected. In order to keep this chapter compact and clear, only the Newtonian case of the finite-width film model will be analyzed here. Changing the aspect ratio has no direct impact on the Deborah number, in contrast to changing the inlet velocity, for instance. Therefore it is not assumed that a viscoelastic study will bring significantly different insight as compared to the Newtonian study together with the previous results of the viscoelastic infinite-width film model.

9.3.1 Combined Effects of Inertia, Gravity and Neck-in

As only the momentum equation is modified by the alternative scaling, the steady state (6.16a–6.16c) remain unchanged, and (6.16d) is modified to

$$\nu_s' = \nu_s \frac{u_s'}{u_s} + F\left(u_s' u_s - 1\right) . \tag{9.23}$$

Similarly, the perturbations (6.17a–6.17c) can still be used and (6.17d) becomes

$$N' = -(H' + L') + \frac{F}{\nu_s}\left[\left(1 - u_s' u_s\right) N + \left(\omega u_s + 2u_s' u_s\right) U + u_s^2 U'\right] . \tag{9.24}$$

As already presented in Sect. 6.3, the Trouton ratio can be used to measure the type of elongational deformation and to determine the parameter region, within which purely uniaxial or purely planar deformation is present. Therefore the Trouton ratio

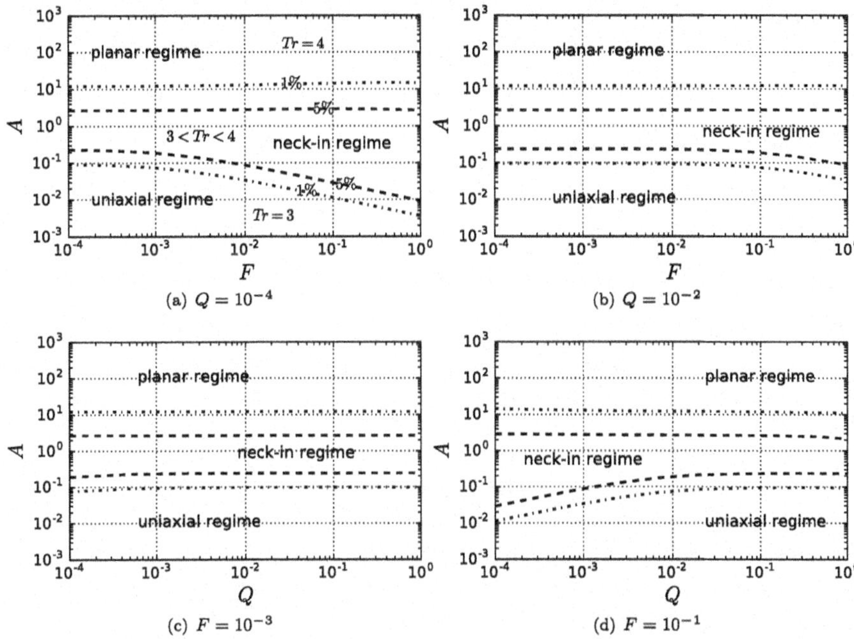

Fig. 9.6 a–b Threshold values of the aspect ratio between the uniaxial and planar regimes and the regime, within which neck-in influences the stability behavior. The inner, dashed lines correspond to a deviation of 5% of the Trouton ratio from the constant value and the outer, dashed-dotted lines correspond to a deviation of 1%, as depicted in **a**. Adapted with permission from Bechert et al. (2016) and AIP Publishing

was calculated with respect to A for various combinations of F and Q and the parameter sets, at which the Trouton ratio differs by 1% and 5% from the constant value, i.e., 3 or 4, were determined. Figure 9.6 visualizes the corresponding threshold lines for constant values of Q in the F–A plane and for constant values of F in the Q–A plane. The parameter regions within which the Trouton ratio is a non-constant variable correspond to the neck-in regime, where the width reduction has an influence on the stability behavior. On the other hand, if the Trouton ratio is constant, the results for the infinite-width film model presented in Sect. 9.2 can be used without any constraints. In the case of the uniaxial regime, however, rescaling of the fluidity by 3/4 is necessary.

The boundary to the planar regime is almost completely independent of F and Q and can be taken as $A = 10$ for practical purpose. In contrast to that, the threshold value of A to the uniaxial regime decreases if Q is small and F is large simultaneously, which becomes visible for $F/Q > 10$. According to (9.11), this means that the effect is more pronounced for large values of Re/Fr and is therefore caused by gravity. Dobroth and Erwin (1986) stated that neck-in is caused by a superposition of planar elongation near the center streamline of the film and uniaxial elongation close to the edges. As gravity acts as a constant force along the main axis, it enforces the

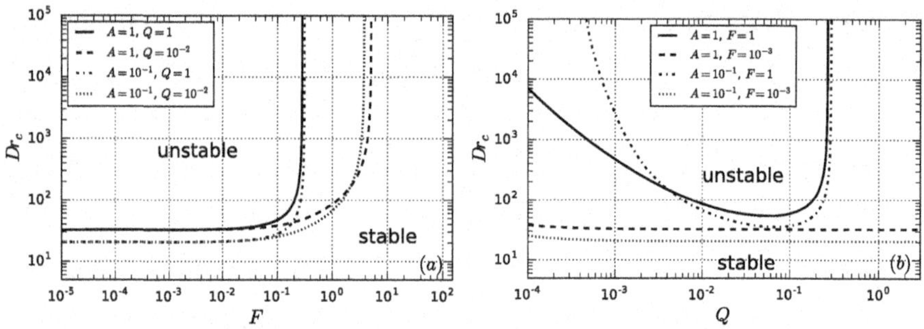

Fig. 9.7 Stability curves: **a** Dr_c versus F for various values of Q and A. **b** Dr_c versus Q for various values of F and A for the finite-width film casting model. Plot **a** reproduced with permission from Bechert et al. (2016) and AIP Publishing

amount of planar deformation and consequently reduces the purely uniaxial regime. For large aspect ratios, however, this effect does not seem to influence the threshold to the planar regime. As the practical range of aspect ratios lies within $0.1 < A < 100$ according to Table 9.1, the uniaxial regime will not be relevant in most cases, in contrast to the threshold to the planar regime at $A = 10$.

Typical neutral stability curves are plotted in Fig. 9.7. The qualitative dependence of the critical draw ratio on the fluidity does not change in comparison with the infinite-width film model (Fig. 9.2). The viscous plateau value for small F is changed due to the neck-in effect as shown in Fig. 6.2a and the point of divergence F_{us} changes slightly for different aspect ratios. The stability minimum with respect to Q, which was already found in the case of infinite width, can be seen for the curves corresponding to $F = 1$ in Fig. 9.7b. Additionally, the plot in Fig. 9.7b reveals that the stabilizing effect of neck-in depends on the value of Q as well, as the two curves for $F = 1$ and different aspect ratios intersect. Similar to the analysis of the infinite-width film model, F_{us} can be determined by fitting the data with a diverging function in the vicinity of the divergence. Figure 9.8 shows the dependence of F_{us} on Q for various values of A, together with the planar and uniaxial limit. For $Q > 0.1$, F_{us} is proportional to $1/Q$, indicating that the threshold to unconditional stability depends on the Reynolds number only. For $Q < 0.1$, however, the shape of the curves varies strongly with A. Especially for $A = 0.1$, F_{us} becomes more than one order of magnitude smaller compared to the planar limit. Similar to the curves shown in Fig. 9.7b, the threshold lines for different aspect ratios can intersect and the minimum value of F_{us} is a function of both A and Q in general. For $A > 10$, the planar limit is recovered. For $A < 0.1$, the uniaxial limit is reached only if $Q > 0.1$, and A needs to be below 10^{-3} to obtain the limit for all values of Q. This can be explained with the results shown in Fig. 9.6, where it can be seen that gravity enlarges the neck-in regime towards small aspect ratios if F/Q is large.

Figure 9.9 shows the stability maps for aspect ratios $A = 0.1$, 1 and 10, analogue to Fig. 9.4a for the infinite-width film model, i.e., $A \to \infty$. It can be seen that the qualitative shape of the isolines of constant draw ratio does not change for different

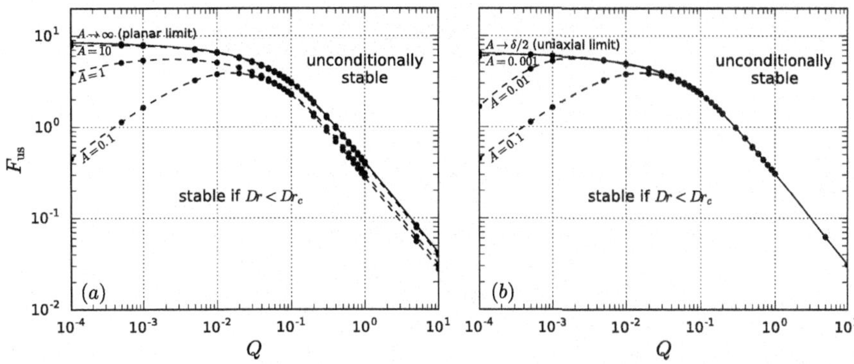

Fig. 9.8 a–b Threshold value F_{us} to unconditional stability for aspect ratios $A = 0.001, 0.01, 0.1, 1, 10$ and the planar and uniaxial limits. For $F > F_{\mathrm{us}}$, no draw resonance occurs for any value of Dr, while for $F < F_{\mathrm{us}}$, the system becomes unstable for $Dr > Dr_{\mathrm{c}}$. The points were determined from the corresponding neutral stability curves Dr_{c} versus F and the dashed lines are approximating splines. Data were split in two plots for the sake of readability. Reproduced with permission from Bechert et al. (2016) and AIP Publishing

aspect ratios within most of the parameter space. The absolute value of Dr_{c} for the viscous regime changes with A due to the stabilizing effect of neck-in, as visualized in Fig. 6.2, and the transition to unconditional stability is modified, as already discussed above. Between these two regions, the critical draw ratio depends on either F/Q or FQ, apart from small regions of transition, as it was already found for $A \to \infty$ and shown in Fig. 9.4b. This reveals that most of the stability dynamics are governed either by viscous, gravity and neck-in, or viscous, inertia and neck-in effects. For this reason, these two limit cases are analyzed separately in the following.

9.3.2 Viscous–Inertia–Neck-in Regime

The *viscous–inertia–neck-in model* is obtained by dropping the gravity term in the momentum equation (9.4). Additionally, the "traditional" scaling transformation, which was used in Sect. 6.2, is applied, such that the boundary conditions become independent of the control parameters and are given by (6.12). The momentum equation then reads

$$F Q h\ell \left(\partial_t u + u \partial_x u\right) = \partial_x \left(h\ell v\right) . \tag{9.25}$$

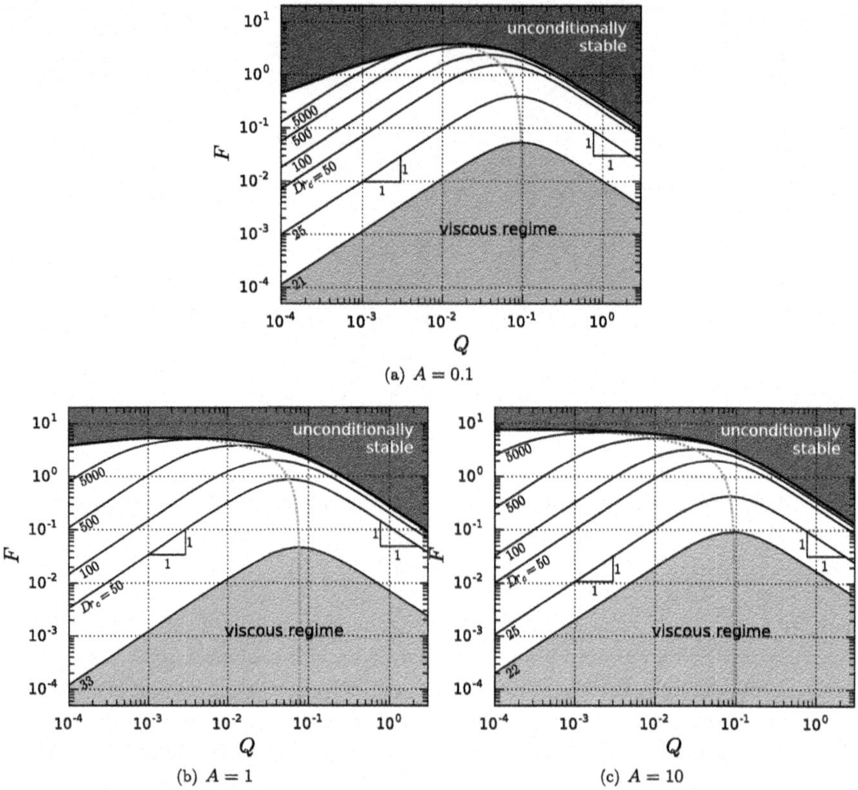

Fig. 9.9 a–c Stability maps for aspect ratios $A = 0.1$, 1 and 10 for the finite-width film model. The thin lines correspond to parameter combinations with constant critical draw ratio, with particular values shown below the lines. The light shaded region belongs to the viscous regime, where the critical draw ratio can be taken as constant. Within the dark shaded region, which is bounded by a thick line, no draw resonance occurs for any value of Dr. The triangle slopes correspond to the logarithmic values. The gray dotted lines indicate stability minima with respect to Q for constant values of F. Adapted with permission from Bechert et al. (2016) and AIP Publishing

The steady state equation for the normal stress difference can be written as

$$\nu_s' = \nu_s \frac{u_s'}{u_s} + F Q u_s u_s' \,, \tag{9.26}$$

the perturbation equation becomes

$$N' = -(H' + L') + FQ\frac{u_s u_s'}{\nu_s}\left[-N + \left(\frac{\omega}{u_s'} + 2\right) U + \frac{u_s}{u_s'} U'\right] , \tag{9.27}$$

while (6.16a–6.16c) and (6.17a–6.17c) remain unchanged.

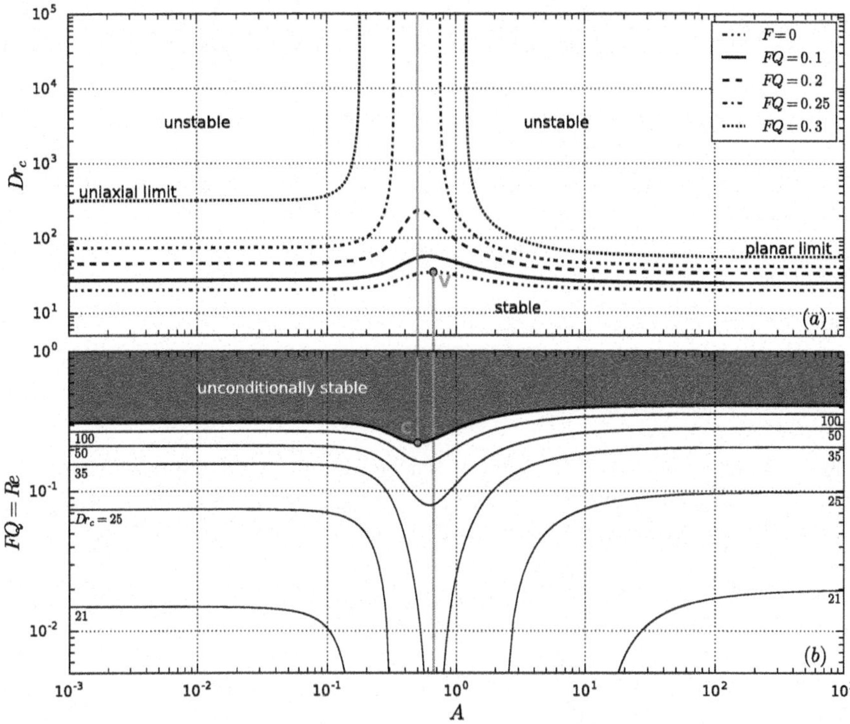

Fig. 9.10 Visualization of the stability behavior of the viscous–inertia–neck-in model. **a** Dependence of the critical draw ratio on the aspect ratio for various values of FQ, which is identical to Re. Label 'V' indicates the stability maximum of the viscous limit, i.e., $F = 0$, which is located at $A = 0.66$ and where $Dr_c = 35.1$. **b** Stability map in the A–FQ plane. The thin lines correspond to parameter combinations with constant critical draw ratio, with particular values shown below the lines. The dark shaded region indicates unconditional stability. Label 'C' indicates the minimum threshold value of FQ to unconditional stability, equal to 0.22 and located at $A = 0.5$. Reproduced with permission from Bechert et al. (2016) and AIP Publishing

Analyzing the stability maps of Fig. 9.9 reveals that the effect of gravity is negligible and thus this model is appropriate for $Q > 0.1$. Figure 9.10 shows the dependence of the critical draw ratio on the aspect ratio for various values of FQ, as well as a stability map with isolines of constant critical draw ratio in the A–FQ plane. The stabilizing influence of inertia over the entire range of aspect ratios is clearly visible. The neutral stability curves of Fig. 9.10a move to larger Dr_c with increasing FQ, i.e., increasing Re. Moreover, the magnitude of the stability maximum, which is located at $A = 0.66$ in the viscous case and denoted by 'V' in the plot, increases with respect to the uniaxial and planar limit, finally leading to an opening of the curves and revealing a window of unconditional stability within the domain of A. This means that the threshold value of FQ to unconditional stability can be lowered by choosing an appropriate aspect ratio. This can be understood with the help of the stability mechanism explained in Sect. 4.2. The stabilizing effect of neck-in increases the threshold value for the effective draw ratio and therefore unconditional stability is reached already for smaller Reynolds numbers. The minimum threshold value is located at $A = 0.5$ and $FQ = 0.22$, as marked in Fig. 9.10b by label 'C'. Consequently, the threshold value varies non-monotonically from the uniaxial limit value of 0.31 to the planar limit value of 0.41, if the aspect ratio is increased. The stability maximum moves to smaller aspect ratios if FQ is increased, but the total difference between the viscous value of $A = 0.66$ and the threshold value of $A = 0.5$ is small. The values of Dr_c in Fig. 9.10a and FQ in Fig. 9.10b differ in the uniaxial and planar limits due to the different Trouton ratios, i.e., 3 and 4, respectively.

9.3.3 Viscous–Gravity–Neck-in Regime

The *viscous–gravity–neck-in regime* is reached for $Q < 0.1$ and if additionally F is not too large, as visible in Fig. 9.9. The corresponding inertia-free momentum equation, using again the traditional scaling, reads

$$-\frac{F}{Q} h\ell = \partial_x (h\ell\nu) , \tag{9.28}$$

leading to the following steady state and perturbation equations for the normal stress difference,

$$\nu_s' = \nu_s \frac{u_s'}{u_s} - \frac{F}{Q} , \tag{9.29}$$

$$N' = -(H' + L') + \frac{F}{Q} \frac{N}{\nu_s} , \tag{9.30}$$

while (6.16a)–(6.16c) and (6.17a)–(6.17c) remain unchanged.

Figure 9.11a depicts the influence of F/Q on the neutral stability curve Dr_c versus A. Similar to inertia, gravity raises the critical draw ratio for the entire domain of aspect ratios. Moreover, the stability maximum is clearly shifted towards smaller aspect ratios with increasing effect of gravity. This can also be seen in Fig. 9.11b, where the isolines of Dr_c are plotted in the A–F/Q plane and the dip of the curves corresponds to the stability maximum. The locus of the maximum in dependence of F/Q was analyzed separately. Figure 9.12 shows the value of Dr_c at the maximum, the ratio of maximal Dr_c to the planar limit value and the aspect ratio at which the maximum is reached. As already observed, the maximal value of Dr_c increases monotonically with increasing F/Q (Fig. 9.12a). Comparison with the planar limit value (Fig. 9.12b) reveals that gravity does not only increase the overall stability, but also the stabilizing influence of neck-in. While neck-in increases the critical draw ratio by a maximum factor of 1.7 in the viscous limit, i.e., $F = 0$, it raises Dr_c by up to one order of magnitude if $F/Q = 10^3$. As can be seen in Fig. 9.12c, the aspect ratio leading to maximal stability lies between 0.1 and 0.66, and is monotonically

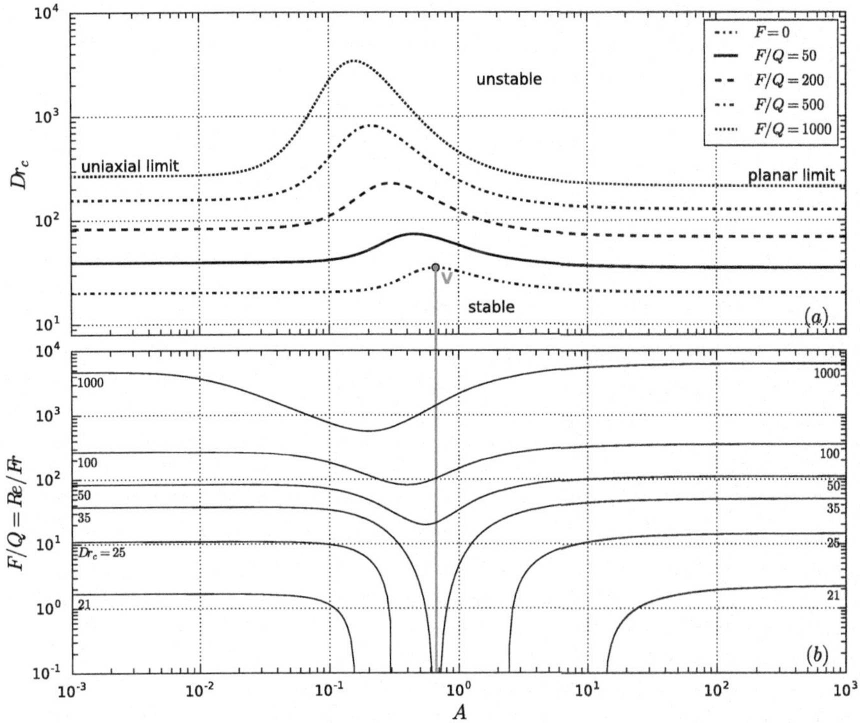

Fig. 9.11 Visualization of the stability behavior of the viscous–gravity–neck-in model. **a** Dependence of the critical draw ratio on the aspect ratio for various values of F/Q, which is identical to Re/Fr. Label 'V' indicates the stability maximum of the viscous limit, i.e., $F = 0$, which is located at $A = 0.66$ and where $Dr_c = 35.1$. **b** Stability map in the A–F/Q plane. The thin lines correspond to parameter combinations with constant critical draw ratio, with particular values shown below the lines. Reproduced with permission from Bechert et al. (2016) and AIP Publishing

decreasing with increasing F/Q. This observation fits to the evolution of the neck-in regime shown in Fig. 9.6. For large values of F/Q, the neck-in regime grows asymmetrically, shifting its center towards smaller aspect ratios.

9.4 Fiber Drawing

A stability analysis of the fiber drawing model derived in Sect. 2.1 includes surface tension as additional effect besides inertia and gravity. Similar to the viscoelastic film study in Sect. 9.2.2, this implies a superposition of both stabilizing and destabilizing effects, which leads most likely to a non-monotic dependence of Dr_c on the control parameters.

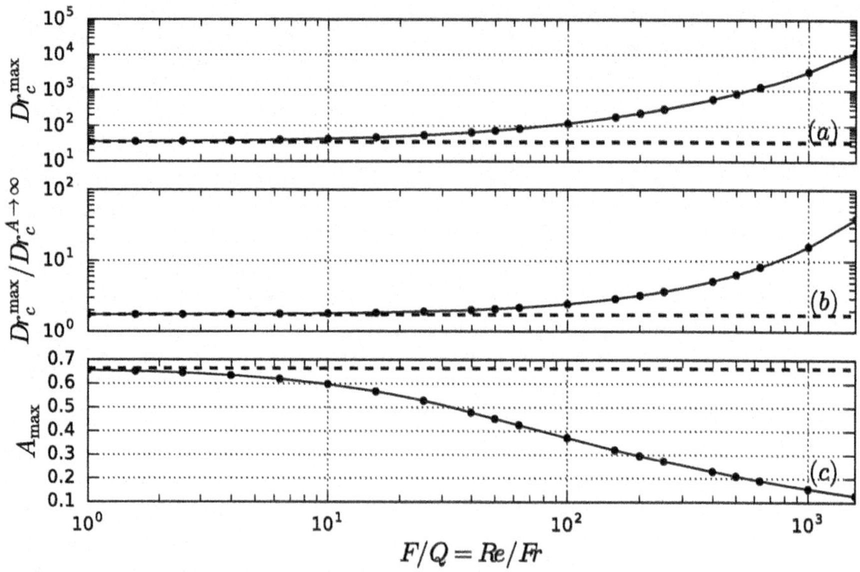

Fig. 9.12 Analysis of the stability maximum of the viscous–gravity–neck-in model with respect to F/Q. The points denote data calculated from the neutral stability curves and the lines are approximated splines. The dashed lines indicate the value of the viscous limit. **a** Critical draw ratio of the stability maximum. **b** Ratio of critical draw ratios of stability maximum to planar limit. **c** Aspect ratio at which the stability maximum is reached. Reproduced with permission from Bechert et al. (2016) and AIP Publishing

9.4.1 Newtonian Model

9.4.1.1 Combined Effects of Inertia, Gravity and Surface Tension

Analogue to Sect. 9.2.1, the normal stress difference (2.18) is directly substituted into the momentum equation (2.24) and the spatial derivative of the velocity is used instead as state variable. The steady state equations then read

$$a_s' = \frac{a_s u_s'}{u_s} \,, \tag{9.31a}$$

$$u_s'' = -\frac{a_s' u_s'}{a_s} + \frac{F}{3}\left(u_s u_s' - 1 - S\frac{a_s'}{a_s^{3/2}}\right) \,, \tag{9.31b}$$

and the perturbation equations are given by

$$A' = -\left(\frac{\omega}{u_s}A + U'\right), \tag{9.32a}$$

$$U'' = \frac{\omega u_s'}{u_s^2}A + \frac{F}{3}\left[\left(\frac{1}{u_s} + \omega + u_s'\right)U + u_s U' + \frac{S}{a_s^{1/2}u_s^2}\left(\left(\omega - \frac{u_s'}{2}\right)A - u_s' U + u_s U'\right)\right]. \tag{9.32b}$$

Typical neutral stability curves for various parameter combinations are visualized in Fig. 9.13. The qualitative shape of the stability curves Dr_c versus F of Fig. 9.13a differs from the previously discussed film casting models for most combinations of S and Q. While the curve corresponding to $S = 10^{-3}$ and $Q = 10^{-1}$ shows diverging behavior and thus a region of unconditional stability, surface tension leads for the other parameter combinations to a region of unconditional instability defined by $Dr_c \leq 1$, if F is above a threshold value. The curve for $S = 10^{-1}$ and $Q = 10^{-3}$ is monotonically decreasing, i.e., increasing the fluidity has a purely destabilizing effect, similar to increasing $1/Ca^*$ in Fig. 5.1a. For $S = 10^{-3}$ and $Q = 10^{-3}$, the stability curve reaches a maximal critical draw ratio at $F \approx 2$. This indicates that for $F < 2$, the stabilizing influence of gravity and/or inertia lead to an increase of Dr_c, while for $F > 2$, the destabilizing effect of surface tension dominates the total stability. Finally, the curve for $S = 10^{-1}$ and $Q = 10^{-1}$ diverges at $F \approx 7$, leading to unconditional stability, but this region is limited to $7 \lesssim F \lesssim 26$, as surface tension effects directly lead to unconditional instability for $F > 26$. The reason for the complex behavior of the neutral stability curves is that while S is a measure of the magnitude of surface tension effects, F increases the magnitude of gravity, inertia and surface tension as compared to viscous forces simultaneously, and therefore tunes both stabilizing and destabilizing factors.

As can be seen in Fig. 9.13b, the dependence of Dr_c on the inlet velocity Q changes as well qualitatively as compared to the film models. For $S = 10^{-3}$ and $F = 10^{-2}$, increasing Q at small values has a destabilizing effect, while increas-

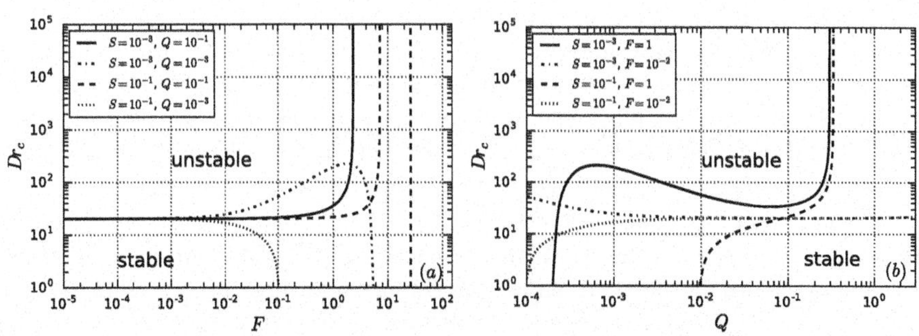

Fig. 9.13 Stability curves **a** Dr_c versus F for various values of Q and S and **b** Dr_c versus Q for various values of F and S for the Newtonian fiber drawing model

ing Q at large values is stabilizing, even though the latter is barely visible for this parameter combination. If, however, $S = 10^{-1}$ instead, increasing Q has a purely stabilizing effect for all values of Q. For $F = 1$, the region of unconditional instability appears for $Q < 10^{-2}$. The curve for $S = 10^{-3}$ and $F = 1$ shows both a stability maximum at $Q \approx 6 \cdot 10^{-4}$ and a stability minimum at $Q \approx 7 \cdot 10^{-2}$, which increases the complexity of the influence of variations in Q on the critical draw ratio.

Figure 9.14 depicts stability maps in the Q–F plane for four values of the surface tension parameter, $S = 10^{-3}$, 10^{-2}, 10^{-1}, and 1. For rather large values of Q, all maps show a similar behavior as compared to the infinite-width film casting model shown in Fig. 9.4, i.e., increasing F has a purely stabilizing effect, a threshold value to unconditional stability exists, and increasing Q has a purely stabilizing effect as well. The threshold to the region of unconditional stability was determined here for reasons of simplicity by tracking the solution branch corresponding to $Dr_c = 10^5$ in the Q–F plane. For rather small inlet velocities, the situation changes significantly. For instance, in Fig. 9.14a large values of F lead to unconditional instability instead of unconditional stability if $Q < 2 \cdot 10^{-3}$. This region is crossed out in the stability maps and limited by a thick line corresponding to $Dr_c = 1$. As already depicted in Fig. 9.13a, the critical draw ratio can depend non-monotonically on F. The dashed line in the stability maps indicates a stability maximum with respect to F for constant Q and therefore divides the parameter space into a region where increasing F has a stabilizing effect and a region where it has a destabilizing effect. For $S = 10^{-1}$ and $S = 1$, the influence of surface tension is strong enough so that increasing F has either a purely stabilizing, or a purely destabilizing effect for a constant value of Q.

As S increases, the region of unconditional instability grows and covers almost half of the practical control parameter space if $S = 1$. The region of unconditional stability shrinks simultaneously and can only be reached for $Q > 0.5$ if $S = 1$. The threshold lines to unconditional stability and instability converge in all cases for large values of F, as well as all other isolines of constant Dr_c. In the vicinity of this line, a small variation of the inlet velocity changes the sign of the divergence of a Dr_c versus F curve, i.e., changes from unconditional stability to unconditional instability. The typical parameter ranges for fiber spinning processes as depicted in Fig. 9.1 show that this dramatic change in stability may indeed be crucial for practical application. As already revealed above, surface tension also affects the stability curves with respect to Q for constant values of F. While Dr_c increases monotonically with Q for $S = 10^{-1}$ and 1, there are stability maxima and minima for $S = 10^{-3}$ and $S = 10^{-2}$, which are visualized by black and, respectively, gray dotted lines. This means that these dotted lines separate the parameter space into regions where increasing Q has a stabilizing effect and regions where it has a destabilizing effect. A local stability maximum, i.e., with respect to both F and Q, is present if the black dotted and dashed lines coincide. Apart from the region of unconditional stability, those parameter combinations are the most recommended for reaching high process stability. For $S = 10^{-1}$ and 1, the viscous regime was defined by $20 < Dr_c < 21$.

9.4.1.2 Viscous–Gravity–Surface-Tension Regime

Similar to the stability maps for the film models, the isolines in Fig. 9.14 show either $F \propto Q$ or $F \propto 1/Q$ within large regions of the parameter space. In the former case, $Dr_c = Dr_c(F/Q)$ for constant values of S, and as $Re/Fr = F/Q$ and $1/Ca^* = 2S(F/Q)$ according to (9.11), a *viscous–gravity–surface-tension model*, neglecting inertia, seems to be sufficient in this region. Applying the traditional scaling (2.5) with boundary conditions (2.25), the momentum equation (9.5) reduces to

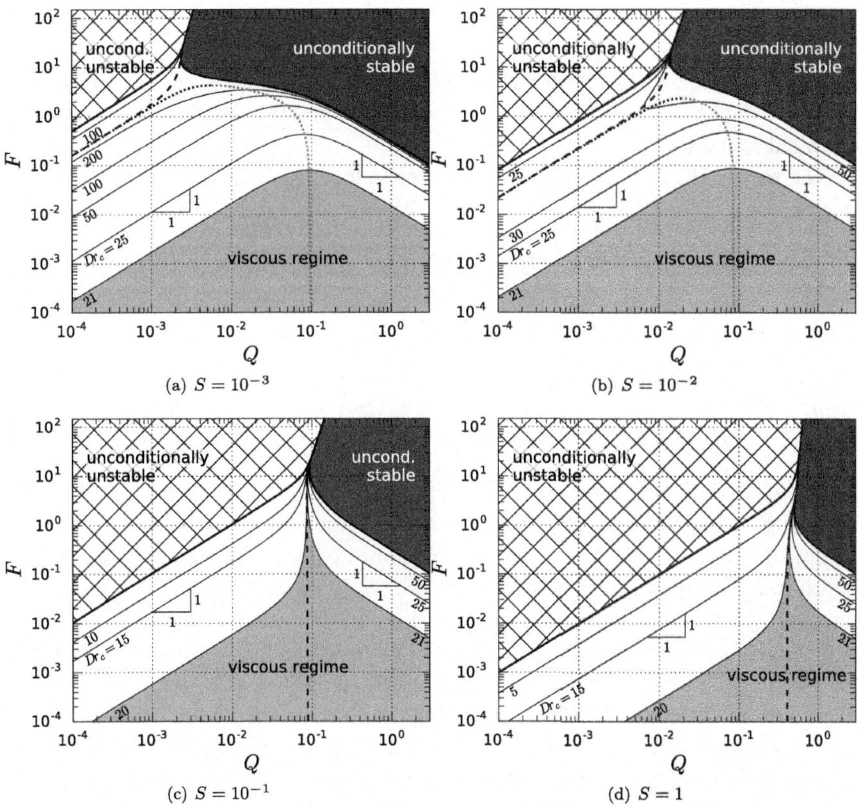

Fig. 9.14 a–d Stability maps for the Newtonian fiber drawing model for dimensionless surface tension parameter $S = 10^{-3}, 10^{-2}, 10^{-1}, 1$. The thin lines correspond to parameter combinations with constant critical draw ratio, with particular values shown below the lines. The light shaded region belongs to the viscous regime, where the critical draw ratio can be taken as constant. Within the dark shaded region, which is bounded by a thick line, no draw resonance occurs for any value of Dr. Within the crossed region, which is bounded by a thick line as well, no stable drawing is possible as $Dr_c \leq 1$ due to surface tension effects. The dotted black (resp. gray) lines indicate maxima (resp. minima) of Dr_c with respect to Q for constant values of F. The dashed line indicates stability maxima with respect to F for constant values of Q. If both types of maxima coincide, a local stability maximum is present. The triangle slopes correspond to the logarithmic values

$$-\frac{F}{Q}\,a\left(1+S\frac{\partial_x a}{a^{3/2}}\right)=\partial_x(av)\,. \tag{9.33}$$

On the other hand, if $Dr_c = Dr_c(FQ)$ for constant values of S, both gravity and surface tension can be neglected, as both effects are tuned by F/Q for constant values of S, as shown by the transformation rules (9.11), and this region refers to a viscous–inertia regime. In fact, within these particular regions the maps are invariant to changes of S, and coincide as well with the corresponding part of Fig. 9.4a if F is scaled accordingly. This reveals that most of the stability dynamics in Newtonian fiber drawing are governed by either viscous–gravity–surface-tension effects, or viscous–inertia effects for most practical parameter combinations. As the results of the viscous–inertia regime for infinite-width film casting presented in Sect. 4.2 can easily be transferred to the fiber drawing model by multiplying the Reynolds number $Re = FQ$ by a factor of 3/4, only the viscous–gravity–surface-tension model is analyzed separately in the following.

The steady state equation (9.31b) reduces in the absence of inertia to

$$u_s'' = -\frac{a_s' u_s'}{a_s} - \frac{F}{3}\left(1+S\frac{a_s'}{a_s^{3/2}}\right), \tag{9.34}$$

and perturbation equation (9.32b) becomes

$$U'' = \frac{\omega u_s'}{u_s^2}A + \frac{F}{3}\left[\frac{U}{u_s} + \frac{S}{\sqrt{a_s}u_s^2}\left(\left(\omega - \frac{u_s'}{2}\right)A - u_s'U + u_sU'\right)\right], \tag{9.35}$$

while (9.31a) and (9.32a) remain unchanged.

In Fig. 9.15a, neutral stability curves Dr_c versus F/Q are shown for various values of S. In the absence of surface tension, i.e., $S = 0$, increasing F/Q has a purely stabilizing effect. Increasing S, however, leads to a region of unconditional instability as the curves drop below $Dr_c = 1$ for F/Q above a threshold value, which itself decreases with increasing S. For moderate values of S, e.g., 10^{-3} and 10^{-2}, the interplay between gravity and surface tension leads to a stability maximum. This maximum is visualized by a dashed line in the stability map depicted in Fig. 9.15b and vanishes for $S > 5.53 \times 10^{-2}$, where increasing F/Q has a purely destabilizing effect. A relevant effect of the stability maximum becomes noticeable for $S < 10^{-2}$, as the largest critical draw ratio is at least 50 then. The threshold to unconditional instability, i.e., the isoline for $Dr_c = 1$, exhibits almost a reciprocal proportionality between S and F/Q, and a rough approximation yields

$$\frac{FS}{Q} \gtrsim 7 \tag{9.36}$$

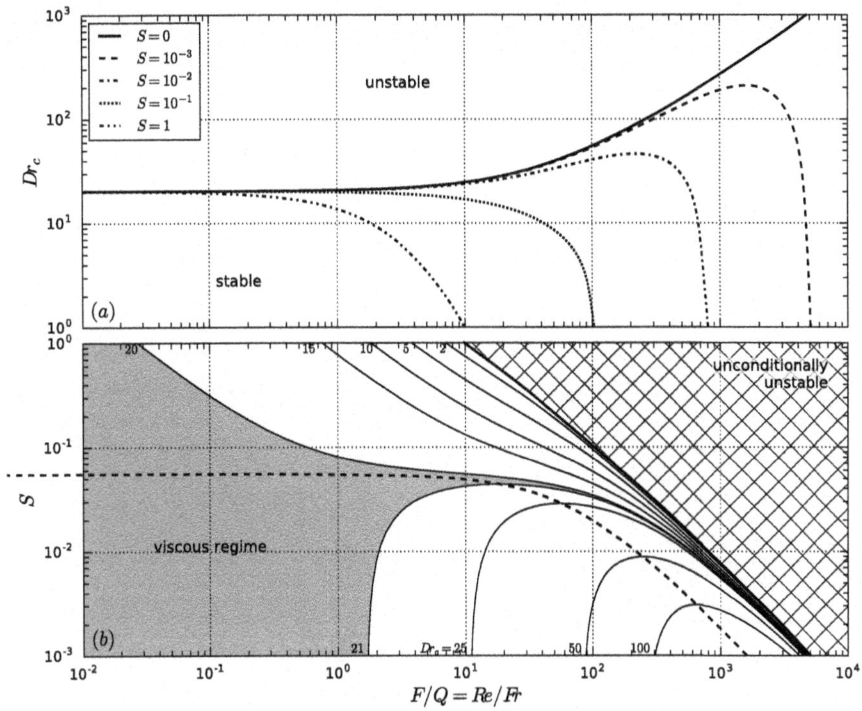

Fig. 9.15 Visualization of the stability behavior of the viscous–gravity–surface-tension model. **a** Dependence of the critical draw ratio on F/Q, which is identical to Re/Fr, for various values of surface tension parameter S. **b** Stability map in the F/Q–S plane. The thin lines correspond to parameter combinations with constant critical draw ratio, with particular values shown to the left of the lines. The thick line indicates the transition to unconditional instability, where $Dr_c \leq 1$. The dashed line indicates maxima of Dr_c with respect to F/Q at constant S and separates the parameter space into regions where increasing F/Q has a purely stabilizing effect (*below*) and where it is purely destabilizing (*above*). The light shaded region marks the viscous regime, where $20 < Dr_c < 21$

as criterion for unconditional instability. As $2FS/Q = 1/Ca^*$, this value is close to the threshold value of 8.13 obtained in Sect. 5.1. The transition to unconditional instability caused by the effect of surface tension is therefore hardly influenced by gravity.

9.4.2 *Viscoelastic Model*

As already noted in Sect. 9.2.2, the PTT constitutive equation differs by more than a global factor for uniaxial and planar elongation. For this reason, a viscoelastic fiber model will be analyzed separately here, even though the dimensionless surface tension parameter S does not influence directly the magnitude of viscoelastic effects, similar to the aspect ratio A. Analogue to the infinite-width film casting model, the nonlinear PTT parameter ε is set to 0.1 and S is fixed to 10^{-2} in addition, so that surface tension effects play a significant role.

Following the steps presented in Sect. 7.3.1, the cross-section-averaged PTT equations for uniaxial deformation can be derived as

$$\Lambda\left(\partial_t\overline{\nu}+u\partial_x\overline{\nu}+\partial_x u\left(\overline{\nu}-3\overline{\tau}_{xx}\right)\right)+e^{\varepsilon\Lambda(3\overline{\tau}_{xx}-2\overline{\nu})}\overline{\nu}=3\partial_x u\ , \tag{9.37a}$$

$$\Lambda\left(\partial_t\overline{\tau}_{xx}+u\partial_x\overline{\tau}_{xx}-2\partial_x u\overline{\tau}_{xx}\right)+e^{\varepsilon\Lambda(3\overline{\tau}_{xx}-2\overline{\nu})}\overline{\tau}_{xx}=2\partial_x u\ . \tag{9.37b}$$

Hereby was used that the trace of the extra stress tensor is equal to $\overline{\tau}_{xx}+2\overline{\tau}_{rr}$, as $\overline{\tau}_{\varphi\varphi}$ is equal to $\overline{\tau}_{rr}$.[4] Together with (2.16) and (9.5), the following steady state equations can be deduced,

$$u_s'=\frac{1}{3\tilde{c}_f-F\Lambda\left(u_s^2+Sa_s^{-1/2}\right)}\left(\frac{\nu_s}{\tilde{c}_{\mathrm{PTT},f}}-F\Lambda u_s\right)\ , \tag{9.38a}$$

$$a_s'=-\frac{a_s u_s'}{u_s}\ , \tag{9.38b}$$

$$\nu_s'=-\nu_s\frac{a_s'}{a_s}+F\left(u_s u_s'-1-S\frac{a_s'}{a_s^{3/2}}\right)\ , \tag{9.38c}$$

$$t_s'=\frac{1}{\Lambda u_s}\left(2u_s'\left(1+\Lambda t_s\right)-\frac{t_s}{\tilde{c}_{\mathrm{PTT},f}}\right)\ , \tag{9.38d}$$

with

$$\tilde{c}_f=1+\Lambda\left(t_s-\frac{2}{3}\nu_s\right)\ , \tag{9.39a}$$

$$\tilde{c}_{\mathrm{PTT},f}=e^{\varepsilon\Lambda(2\nu_s-3t_s)}\ . \tag{9.39b}$$

Analogue to the infinite-width film casting model in Sect. 9.2.2, the unscaled perturbation ansatz (9.21) is used for the stress components in order to increase the numerical stability. This leads to the following perturbation equations,

[4] Both $\overline{\tau}_{\varphi\varphi}$ and $\overline{\tau}_{rr}$ obey the same differential equation and are identical in the Newtonian limit.

$$U' = \frac{1}{u_s\left(3\tilde{c}_f - F\Lambda\left(u_s^2 + Sa_s^{-1/2}\right)\right)}\left[\left(\frac{1-2\varepsilon\Lambda v_s}{\tilde{c}_{\mathrm{PTT},f}} + \Lambda(\omega + 2u_s')\right)\tilde{N}\right.$$
$$+ \Lambda\left(v_s(u_s' + \omega) - u_s v_s' + F\left(u_s(u_s u_s' - 1) + \frac{S}{2\sqrt{a_s}}(2\omega + u_s')\right)\right)H \tag{9.40a}$$
$$\left. - \left(3\tilde{c}_f u_s' + \Lambda\left(v_s u_s' - u_s v_s' - F u_s^2(\omega + 2u_s')\right)\right)U + 3\Lambda\left(\frac{\varepsilon v_s}{\tilde{c}_{\mathrm{PTT},f}} - u_s'\right)\tilde{T}\right],$$

$$A' = -\left(\frac{\omega}{u_s}A + U'\right), \tag{9.40b}$$

$$\tilde{N}' = -v_s A' - \frac{a_s'}{a_s}\tilde{N} + F\left[u_s^2 U' + u_s(\omega + 2u_s')U + \frac{S}{\sqrt{a_s}}\left(\frac{a_s'}{2a_s}A - A'\right)\right], \tag{9.40c}$$

$$\tilde{T}' = \frac{1}{u_s\Lambda}\left[2u_s(1+\Lambda t_s)U' + \frac{t_s}{\tilde{c}_{\mathrm{PTT},f}}U + \frac{2\varepsilon\Lambda t_s}{\tilde{c}_{\mathrm{PTT},f}}\tilde{N}\right.$$
$$\left. - \left(\frac{1+3\varepsilon\Lambda t_s}{\tilde{c}_{\mathrm{PTT},f}} + \Lambda(\omega - 2u_s')\right)\tilde{T}\right]. \tag{9.40d}$$

As already observed for the viscoelastic film casting model, the numerical continuation fails for large values of the fluidity, which is probably due to the strong gradients in the steady states. For this reason, the analysis will again be limited to $F < 1$, which rather excludes the large scale industrial processes, as Fig. 9.1 shows. This leads to two important conclusions. On the one hand, the fact that this particular parameter region seems to be accessible only for almost negligible values of Λ indicates that materials for industrial, high-speed application should be selected to have a small relaxation time compared to the process time, i.e., a Deborah number of $\mathcal{O}(10^{-3})$ or lower, approaching the Newtonian limit. On the other hand, this also reveals that besides the stability analyses presented within this book, steady state analyses should be performed as well, which define the available parameter space. For this, conditions for realizable steady states need to be set beforehand. As such an analysis is almost as extensive as the stability analysis, and as such conditions additionally depend on the particular process setup, it will be concentrated here on the reduced parameter space, which still covers the small-scale fiber spinning processes (see Table 9.1), and which is accessible by the implemented numerical method.

Stability maps in the Q–F space for four different values of Λ are depicted in Fig. 9.16. Similar to the viscoelastic film casting scenario, two major regions can be identified. If the isolines are parallel to the F-axis, only viscoelastic effects influence the stability. With increasing Λ, this region grows towards small values of Q. Especially for $\Lambda = 0.1$ and 1, a viscous–gravity–surface-tension region is visible for small values of Q, including a region of unconditional instability and a local maximum with respect to both F and Q. If Λ becomes larger, viscoelastic effects push this region out of the relevant parameter space, and the maxima with respect to F and Q start to deviate from each other, until the one with respect

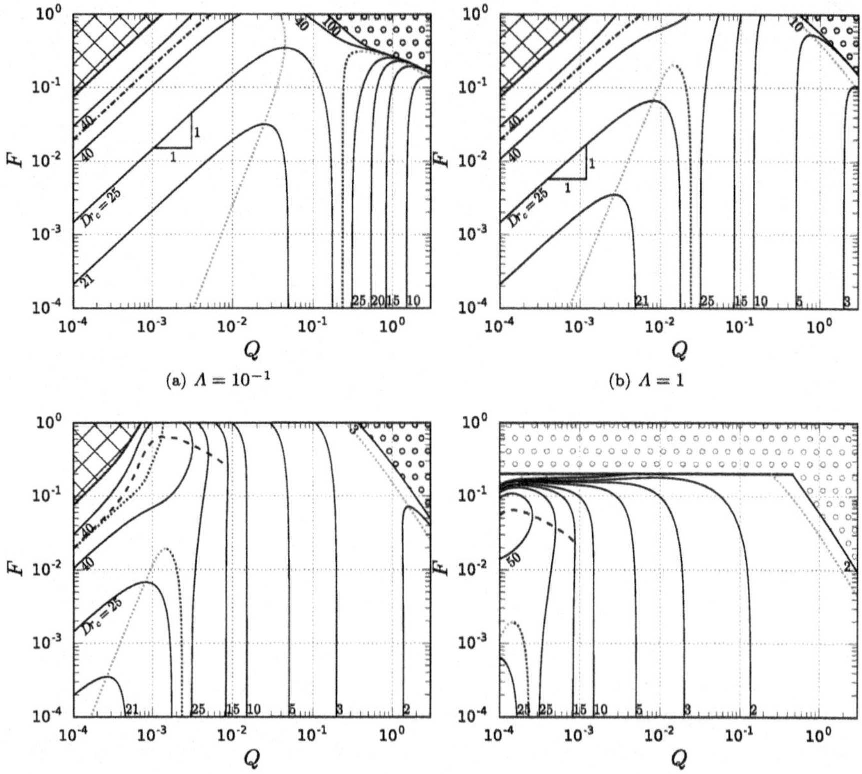

(a) $\Lambda = 10^{-1}$ (b) $\Lambda = 1$

(c) $\Lambda = 10$ (d) $\Lambda = 10^2$

Fig. 9.16 a–d Stability maps for dimensionless relaxation times $\Lambda = 10^{-1}, 1, 10, 10^2$ for the PTT fiber drawing model. The thin lines correspond to parameter combinations with constant critical draw ratio, with particular values shown below or to the right of the lines. Within the dotted region, no solution could be calculated, probably due to high gradients in the steady states. Within the crossed region, which is bounded by a thick line, no stable drawing is possible as $Dr_c \leq 1$ due to surface tension. The black (gray) dotted lines indicate maxima (minima) of Dr_c with respect to Q for constant values of F. The black (resp. gray) dashed lines indicate stability maxima (resp. minima) with respect to F for constant values of Q. If both types of maxima coincide, a local stability maximum is present. The triangle slopes correspond to the logarithmic values

to Q finally disappears. In contrast to the film casting model, no minimum with respect to $De = \Lambda Q$ induced by viscoelastic effects could be observed for uniaxial deformation. The minimum with respect to Q, which can be tracked up to $F = 10^{-4}$, corresponds to the one appearing as well in the Newtonian case (see Fig. 9.14). The maximum close to $Q\Lambda = 0.02$, in contrast, can be related to viscoelasticity, similar to the film casting scenario. The isolines clearly indicate that inertia does not play a major role in the attainable parameter space if viscoelastic effects are strong enough. For moderate values of F, viscoelastic effects suppress the influence of inertia, while for larger values of F, no steady state solutions can be found.

9.5 Choosing the Appropriate Stability Map for a Particular Process

The results presented in this chapter serve primarily the purpose of a direct application in the scope of a particular process. The new scaling based on a fixed cast length L offers dimensionless parameters directly correlated to practical application. Once these parameters calculated for a particular process, the findings presented here enable a deduction of suitable model assumptions, like the importance of neck-in, inertia, gravity or surface tension. Stability maps (Figs. 9.4, 9.5, 9.9, 9.14 and 9.16) with isolines of the critical draw ratio reveal various dynamical regimes. The following general rules can be applied for choosing appropriate model assumptions for a particular process:

- neck-in effects in film casting become important for $A < 10$,
- viscoelastic effects can always be neglected for $\Lambda Q = De < 10^{-3}$,
- isolines with $F \propto Q$ indicate a viscous–gravity(–neck-in/–surface-tension) regime, i.e., negligible inertia and viscoelastic effects,
- isolines with $F \propto 1/Q$ indicate a viscous–inertia(–neck-in) regime, i.e., negligible gravity, surface tension and viscoelastic effects,
- isolines parallel to the F-axis indicate a purely viscoelastic(–neck-in) regime, i.e., negligible inertia, gravity and surface-tension effects.

Gravity and inertia are found to be hardly important simultaneously, as well as surface tension and inertia effects do mostly not superimpose in fiber spinning applications. A general stability analysis within a regime where viscoelastic and inertia effects both play a role appeared to be impossible, as the steady state solutions already restrict the available parameter space. Special trajectories, like the threshold to unconditional (in-)stability or local stability maxima and minima, were analyzed in particular, as they are of high practical importance.

If only one additional effect remains important for the calculation of the critical draw ratio, the simple empirical correlations already proposed in Chap. 4 can be directly applied. In the cases of fiber spinning with non-negligible surface tension and Newtonian finite-width film casting, empirical correlations for the various dynamical regimes were explicitly given as well in Chaps. 5 and 6, respectively. Apart from that, stability maps of all relevant cases presented in this chapter help to estimate the critical draw ratio with respect to the control parameters.

Chapter 10
Non-isothermal Film Casting

Abstract We develop in this chapter how the draw resonance instability is affected by cooling through the temperature dependence of the viscosity. We first discuss the difference between isothermal and non-isothermal draw resonance in literature as well as between non-isothermal film casting and fiber drawing. Next, a non-isothermal model is derived for film casting and a linear stability analysis allows to investigate the influence of the heat transfer on draw resonance and how cooling mainly has a stabilizing effect. The prescribed temperature regime in the limit of infinite Stanton number is also identified and a scenario for the stabilizing mechanism of draw resonance due to cooling is proposed, based on phase shifts between axial tension and relevant spatially averaged quantities of the system. Finally, we test both the Jung's and Kim's stability criteria and show how the different times involved in those criteria are affected by cooling.

10.1 Heat Transfer in Draw Resonance

Despite the low critical draw ratio of $Dr_c = 20.218$ for purely viscous isothermal case, fibers for instance can be stably drawn in industrial processes at much higher speeds, e.g., $Dr = \mathcal{O}(10^5)$. Shah and Pearson (1972a) found that convective heat transfer is responsible for such a strong stabilization. However, as shown by these authors and subsequent works (see e.g., Gupta et al. (1996)), regimes in fiber drawing intrinsically correspond to small Stanton numbers ($St \leq 1$)—the Stanton number, as defined later, is the ratio of the surface heat transfer relative to the bulk energy advection. In contrast, film stretching processes can in certain circumstances correspond to large Stanton numbers ($St \gg 1$). Again from the fiber drawing literature, Fisher and Denn (1977) have shown that though the cooling by convective heat transfer has a stabilizing effect for small Stanton numbers, there should be a reverse trend toward large Stanton numbers. Indeed, the limit $St \to \infty$ means the liquid has obtained the ambient temperature and is effectively in isothermal conditions. However, this is rigorously true if (i) the ambient temperature is constant, which is never the case in practice, (ii) the one-dimensional assumption is applicable

© Springer Nature Switzerland AG 2025

M. Bechert and B. Scheid, *Draw Resonance*, Fluid Mechanics and Its Applications 141,
https://doi.org/10.1007/978-3-031-85568-9_10

(i.e. no radial temperature variations), which strictly forbids the study of large Stanton number ($St \gg 1$) regimes. We also mention that Geyling and Homsy (1980) and Willien et al. (1988) have shown for glass fibers that though pure convective heat transfer is always stabilizing, adding radiative heat transfer can be destabilizing for sufficiently large values of the radiative heat transfer coefficient. This result indicates the non-monotonic behavior of the neutral stability for increasing Stanton number when it depends on both convective and radiative heat transfer coefficients.

In practice, heat transfer implies the transition to a solid, which will simply be modeled here by a drastic increase of the viscosity as the temperature approaches the glass transition temperature. Though this approach ignores details of solidification such as anisotropy depending on the processing conditions in the case of polymer crystallization (see e.g., Barot and Rao (2004) for details), it is expected to be satisfactory in accounting for glass transition. In any case, assuming that heat released during solidification can be ignored is valid because the glass transition occurs at take-up, after the geometry is no longer changing (see e.g., Smith and Stolle (2000a)). In the case of glass fibers, Forest et al. (2000) have considered a more general two-phase model with a solidification front at the transition temperature. They nevertheless reached the conclusion that a 'one-phase' model has the crucial advantage that solutions have continuous gradients at the glass transition free boundary, which not only is consistent with experiments but further enables one to perform a linear stability analysis of these steady boundary-value states; it is a 'one-phase' approach that we thus adopt herein.

Note that contrarily to the purely viscous isothermal case, the temperature equation for film casting and fiber drawing are different, such as only film casting is treated in this chapter and non-isothermal fiber drawing will be treated in Chap. 11.

10.2 Non-isothermal Model

The film geometry given in Fig. 2.1 is adapted to non-isothermal case in Fig. 10.1, where the temperature field $T(x, y, t)$, the inlet temperature T_0 and the ambient far-field temperature $T_a(x)$ have been added.

10.2.1 Flow Field

Relying on the film model developed in Sect. 2.2, the mass conservation equation remains unchanged,

$$\partial_t h + \partial_x (h u^{(0)}) = 0 \,, \tag{10.1}$$

whereas the momentum equation needs a specific attention since the viscosity now depends on temperature, i.e. $\eta(T)$, hence on both coordinates x and y plus t. Using the

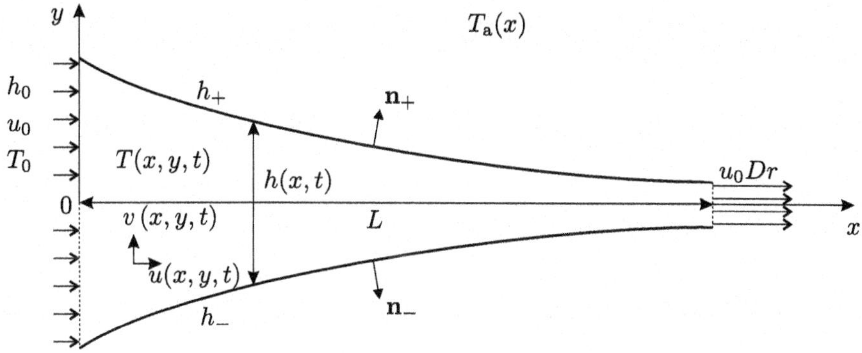

Fig. 10.1 Sketch of non-isothermal film casting problem with dimensional variables and parameters

inlet viscosity $\eta_0 = \eta(T_0)$ as reference and then making the viscosity dimensionless using the following scaling rule

$$\eta \rightarrow \eta_0 \eta \,, \tag{10.2}$$

the axial stress balance (2.34) can be straightforwardly adapted to

$$\sigma_{xx}^{(0)} = 4\eta \partial_x u^{(0)} \,. \tag{10.3}$$

Focusing on the sole effect of non-constant viscosity, inertia and gravity effects are neglected in this chapter. Setting thus $Re = 0$ and $Re/Fr = 0$, and recalling the asymptotic expansion for the shear stress $\sigma_{xy} = \sigma_{xy}^{(0)} + \beta^2 \sigma_{xy}^{(1)} + \mathcal{O}(\beta^4)$, the β^2-order of (2.30a) and (2.31a) reduce to

$$\partial_y \sigma_{xy}^{*(1)} = -\partial_x \sigma_{xx}^{(0)} \,, \tag{10.4}$$

$$\sigma_{xy}^{*(1)}\big|_{h_\pm} = \sigma_{xx}^{(0)}\big|_{h_\pm} \partial_x h_\pm \,, \tag{10.5}$$

where $\sigma_{xy}^* := \beta \sigma_{xy}$. The integration of (10.4) between the lower and upper surfaces, using (10.5), then yields

$$\partial_x \left[\int_{h_-}^{h_+} \sigma_{xx}^{(0)} \, dy \right] = 0, \tag{10.6}$$

which indicates that the depth-averaged tension remains constant along the film. An equivalent constant-tension condition has been derived for fibers by Dewynne et al. (1992). Substituting (10.3) into (10.6), and integrating with respect to x, finally yields

$$\hat{\eta} \, h \, \partial_x u^{(0)} = f, \tag{10.7}$$

where $\hat{\eta} = (1/h) \int_{h_-}^{h_+} \eta \, dy$ is the depth-averaged dimensionless viscosity and $f(t)$ remains the unknown integration constant proportional to the film tension. Equation (10.7) extends the classical equation for an extensional flow of a Newtonian fluid (e.g., see Pearson (1985)) to the case of in-depth variations of the viscosity, which remains in turn to be determined through the temperature field. The system (10.1, 10.7) for the three unknowns $h, u^{(0)}$ and f should be solved with the unchanged boundary conditions,

$$h(0,t) = 1 \,, \quad u^{(0)}(0,t) = 1 \quad \text{and} \quad u^{(0)}(1,t) = Dr. \tag{10.8}$$

10.2.2 Temperature Field

The temperature field in the film is described using the conservation of energy Eq. (1.6),

$$\rho c_p \left(\partial_t T + \mathbf{v} \cdot \nabla T\right) = k \nabla^2 T, \tag{10.9}$$

where all properties are assumed to be independent of the temperature. At the free surfaces, we impose the classical Newton's cooling law,

$$- k \, \mathbf{n}_{\pm} \cdot \nabla T\big|_{h_\pm} = \alpha \left(T\big|_{h_\pm} - T_{\mathrm{a}}(x) \right), \tag{10.10}$$

where α is the heat transfer coefficient, identical and constant on the two interfaces, and $T_{\mathrm{a}} = T_{\mathrm{a}}(x)$ is the prescribed far-field ambient temperature, which for generality we assume may vary along the film. The temperature is non-dimensionalized with the following scaling rules

$$T \to T_0 \, T \quad \text{and} \quad T_{\mathrm{a}} \to T_0 \, T_{\mathrm{a}}. \tag{10.11}$$

Using the other necessary scaling rules from (2.28), (10.9) thus takes the following dimensionless form

$$\beta \, Pe \left(\partial_t T + u \partial_x T + v \partial_y T \right) = \beta^2 \partial_{xx} T + \partial_{yy} T. \tag{10.12}$$

The Péclet number $Pe = \rho c_p u_0 h_0 / k$ measures the rate of convection of energy $\rho c_p u_0$ in the flow direction relative to the rate of heat conduction k/h_0 in the transverse direction. In turn, the boundary condition (10.10) has the dimensionless form

$$\left(1 + \beta^2 \frac{1}{4} (\partial_x h)^2 \right)^{-1/2} \left(\partial_y T\big|_{h_\pm} - \beta^2 \partial_x h_\pm \, \partial_x T\big|_{h_\pm} \right) = \mp Bi \left(T\big|_{h_\pm} - T_{\mathrm{a}} \right), \tag{10.13}$$

where the Biot number $Bi = \alpha h_0/k$ describes the rate of heat transport from the liquid to the surrounding. In this study, we shall consider large Péclet numbers[1] such that $\beta\,Pe = \mathcal{O}(1)$. At leading order, i.e. $\beta^2 \to 0$, (10.12) and (10.13) then reduce to

$$\partial_{yy} T^{(0)} = \beta\,Pe \left(\partial_t T^{(0)} + u^{(0)}\,\partial_x T^{(0)} + v^{(0)}\,\partial_y T^{(0)}\right), \tag{10.14a}$$

$$\partial_y T^{(0)}\big|_{h_\pm} = \mp Bi \left(T^{(0)}\big|_{h_\pm} - T_\mathrm{a}\right). \tag{10.14b}$$

In order to eliminate the transverse coordinate y in the system above, we assume a parabolic temperature profile that satisfies the boundary conditions (10.14b) and explicitly introduces the average temperature $\theta = \theta(x,t) \equiv (1/h)\int_{h_-}^{h_+} T^{(0)}\,dy$. Thus we can write

$$T^{(0)}(x,y,t) = T_\mathrm{a}(x) + \frac{\theta(x,t) - T_\mathrm{a}(x)}{1 + \frac{Bi}{6} h(x,t)} \left(1 + Bi\; h(x,t)\left(\frac{1}{4} - \frac{y^2}{h(x,t)^2}\right)\right). \tag{10.15}$$

Substituting $v^{(0)} = -y\,\partial_x v^{(0)}$ obtained from the continuity Eq. (2.29) expanded at leading order, replacing (10.15) into (10.14a), and integrating the latter between the lower and upper surfaces yield, using (10.1),

$$\partial_t\theta + u\,\partial_x\theta = -St\frac{\theta - T_\mathrm{a}(x)}{h(1 + \frac{Bi}{6}h)}, \tag{10.16}$$

where the superscript "(0)" for u will be omitted from now on. The Stanton number is defined as

$$St = \frac{2Bi}{\beta\,Pe} = \frac{2\alpha L}{\rho c_p u_0 h_0}, \tag{10.17}$$

and compares the actual heat transfer across both interfaces, hence the factor 2, to the energy advected by the main flow. Equation (10.16) needs to be solved with the following boundary condition

$$\theta(0,t) = 1\,. \tag{10.18}$$

In the limit of $St \to 0$, the film retains all of its thermal energy along the stretching direction and the average temperature remains constant everywhere, i.e. $\theta = 1$. This situation occurs for two particular cases: (i) $Pe \to \infty$ which is the case of a poorly conducting fluid and/or relatively fast speeds, though $T^{(0)}$ can still depend on y for finite Bi; (ii) $Bi \to 0$ in the case of no heat loss at the interfaces, i.e. $T^{(0)} = 1$ everywhere. In contrast, the limit $St \to \infty$, when Bi remains finite, means that the typical speed is slow enough and possibly the convective heat transfer high enough, such that the liquid obtains the far-field ambient temperature, i.e. $T^{(0)} = T_\mathrm{a}$.

[1] Considering large Péclet numbers enforce the treatment of the heat convection term (i.e. left-hand-side of (10.12)), which goes beyond the usual approximation of a purely conductive film obtained in the limit of $\beta \to 0$.

In the case now of small heat transfer, $Bi = \mathcal{O}(\beta)$, the temperature profile across the film tends to be flat, i.e. $T^{(0)} = \theta(x, t)$ (as shown by 10.15). If in addition, we set the speed to be moderate such that $St = \mathcal{O}(1)$, (10.16) reduces to

$$\partial_t \theta + u \partial_x \theta = -St \frac{\theta - T_a}{h}. \tag{10.19}$$

This form of the thermal energy equation was first derived by Shah and Pearson (1972a) in the specific case of $T_a = 0$ and for fiber drawing,[2] by solving (10.12) and (10.13) with the assumption $Bi \sim \beta Pe = \mathcal{O}(\beta)$. Consequently, the temperature is found at leading order to be uniform across the thickness in this case and the balance in (10.19) is found at next order. Equation (10.19) thus corresponds to the one-dimensional assumption for the temperature field, which is valid rigorously in the limited range of $St = \mathcal{O}(1)$. It is worth noting that the first order balance (10.19) can be recovered from the leading order one (10.16) by taking the limit $Bi \to 0$. Therefore the parameter Bi in (10.16) appears as a measure of the departure from the one-dimensional assumption for the thermal energy equation. Indeed, (10.16) has been written with no restriction on neither Bi nor Pe, which enables to explore the entire range of Stanton numbers. Note that for very large Pe, inertia effects might have to be considered, as done by Shah and Pearson (1972b) for fiber drawing and by Cao et al. (2005) for film casting.[3]

10.2.3 Depth-Averaged Viscosity

The momentum and thermal energy equations are linked via the viscosity $\eta(T)$. We shall base our model for viscosity variations on the assumptions that no freezing occurs and that the fluid viscosity obeys the Arrhenius equation for molecular kinetics. In dimensionless form, we have

$$\eta(T) = \mathrm{e}^{\mu\,(1/T - 1)}, \tag{10.20}$$

where μ measures the sensitivity of the viscosity with the temperature. From (10.7), we need the average viscosity over the thickness, which is essentially non-algebraic in y for a parabolic temperature profile $T^{(0)}$ as defined in the previous section. To proceed analytically, we then have to make an additional assumption on the temperature field, namely that its variation across the film remains small compared

[2] Note that the general Eq. (10.16) applies for fiber drawing provided h is substituted by $\sqrt{a}$ and the factor 6 by 4. It can be solved together with the flow equations (10.1, 10.7) in which h is substituted by a. This is the reason why the equivalence between the film model and the fiber model does not hold in the non-isothermal case.

[3] Taking $Pe = Pr\,Re$ with $Pr = \mu c_p/k$ the Prandtl number, which is usually much larger than 10^3 for molten glass or polymers, Re should remains small.

to the variations along the film, i.e. $|T^{(0)} - \theta| \ll 1$. This step allows us to average (10.20) across the thickness and truncate the result at first order in $T^{(0)} - \theta$ to obtain

$$\hat{\eta}(\theta) \approx e^{\mu\,(1/\theta - 1)}. \qquad (10.21)$$

We note the fact that $\eta(T)$ and $\hat{\eta}(\theta)$ have the same functional form is fortuitous and is true only at $\mathcal{O}(T - \theta)$ and in the specific case of a symmetric thermal profile. The latter feature would obviously not be the case for different heat transfer coefficients and/or different far-field temperatures at each interface, as it is obviously the case in glass film processes.

10.2.4 Far-Field Ambient Temperature

Practical applications of stretching viscous films might involve heat transfer by convection and/or radiation, which we simulate by using a non-constant ambient temperature, which can mimic in turn an imposed heat flux distribution, namely $Bi T_a(x)$ since the Biot number is constant. We follow here the same approach as used by Willien et al. (1988) for fiber drawing and assume the far-field temperature to decrease exponentially along the stretching direction,

$$T_a(x) = e^{-b\,x}, \qquad (10.22)$$

where b is a positive parameter. Note that $T_a(0) = 1$, which sets the dimensional far-field temperature equal to the surface temperature T_0 at inlet. This is necessary to ensure $|T^{(0)} - \theta| \ll 1$ at the inlet, regardless of the possibility of large Bi. Note some authors (e.g., Shah and Pearson (1972a) or Gupta et al. (1996)) have instead assumed that the heat transfer coefficient depends on the fluid velocity and hence increases along the stretching direction, using a correlation given by Kase and Matsuo (1965). Nevertheless, this has a significant influence only for large draw ratios, i.e. in a small region of the parameter space, and we therefore chose to keep the heat transfer coefficient α constant for the sake of simplicity.[4] In fact, we shall show below that for $St \gg 1$, the critical draw ratio is as small as in isothermal conditions ($St = 0$) and the dependence of α with u can safely be neglected (see Sect. 10.5).

[4] Non-constant heat transfer coefficient will be treated in Chap. 11, considering $St \ll 1$ in fiber drawing.

10.3 Stability Problem

Following the same line as in Sect. 3.2, let us analyze the linear stability of the non-isothermal model by writing the variables as

$$u(x,t) = u_s(x)\left(1 + U(x)\,\mathrm{e}^{\omega t}\right), \tag{10.23a}$$
$$h(x,t) = h_s(x)\left(1 + H(x)\,\mathrm{e}^{\omega t}\right), \tag{10.23b}$$
$$\theta(x,t) = \theta_s(x)\left(1 + \Theta(x)\,\mathrm{e}^{\omega t}\right), \tag{10.23c}$$
$$f(t) = f_s\left(1 + F\,\mathrm{e}^{\omega t}\right), \tag{10.23d}$$

where the subscript 's' identifies again the stationary solutions; U, H, Θ and F representing the perturbations (only the three first are complex) and ω being the eigenvalue. The system of equations to be solved can thus be recast into a set of ODEs for the base state and the perturbations.

10.3.1 Base State

Setting $\partial_t h = 0$ in (10.1), together with the boundary condition (10.8), leads to

$$h_s = \frac{1}{u_s}, \tag{10.24}$$

while the stress and temperature equations (10.7) and (10.16) for the base state are

$$u_s' = \frac{f_s u_s}{\hat{\eta}(\theta_s)}, \tag{10.25a}$$
$$\theta_s' = -St\frac{\theta_s - T_\mathrm{a}}{1 + \frac{Bi}{6}h_s}. \tag{10.25b}$$

These ODEs are to be solved with $u_s(0) = \theta_s(0) = 1$. Notice that if we neglect the thickness variation of the temperature (by setting $Bi \to 0$), (10.25b) can be integrated using (10.22) to give

$$\theta_s(x) = \frac{b\,\mathrm{e}^{-St\,x} - St\,\mathrm{e}^{-bx}}{b - St},$$

which further reduces to

$$\theta_s = \mathrm{e}^{-St\,x} \quad \text{for} \quad T_\mathrm{a} = 0, \tag{10.26a}$$
$$\theta_s = 1 \quad \text{for} \quad St = 0, \tag{10.26b}$$
$$\theta_s = T_\mathrm{a} \quad \text{for} \quad St \to \infty. \tag{10.26c}$$

10.3.2 Perturbation Equations

Substituting (10.23) into (10.1), (10.7) and (10.16), while using the stationary solutions (10.24–10.25) for simplifications, yields three coupled ODEs

$$H' = -U' - \omega h_s H, \tag{10.27a}$$

$$U' = \frac{f_s}{\hat{\eta}(\theta_s)}\left(F - U - H + \mu\frac{\Theta}{\theta_s}\right), \tag{10.27b}$$

$$\Theta' = -\omega h_s \Theta - \frac{St}{\theta_s(1+\frac{Bi}{6}h_s)}\left[T_{\mathrm{a}}\Theta - (\theta_s - T_{\mathrm{a}})\left(U + \frac{1+\frac{Bi}{3}h_s}{1+\frac{Bi}{6}h_s}H\right)\right], \tag{10.27c}$$

which are to be solved with the boundary conditions

$$H(0) = U(0) = \Theta(0) = U(1) = 0\,. \tag{10.28}$$

The amplitude F remains undetermined since the eigenvalue problem (10.27) is linear and homogeneous. This feature means the other amplitudes should be interpreted relative to F. Again, we arbitrarily fix $F = 0.1$ for plotting reason in the following (see also Sect. 3.1).

10.4 Neutral Stability

As performed in the previous chapters, we focus in this section on the neutral stability of the system that separates stable and unstable regions in the parameter space formed by Dr, St, Bi, μ and b. Thus we find solutions to the eigenvalue problem (10.25,10.27,10.28), considering only the first mode of instability, i.e. the largest eigenvalue, when its real part is zero, namely $\omega_R = 0$. As outlined in Sect. 3.2.2, solutions to the problem have been sought by continuation using the software AUTO-07P, starting from the analytical solutions (3.15,3.17) for the isothermal case, i.e. for $St = 0$, with the first mode (3.19) corresponding to $Dr_{\mathrm{c}} = 20.218$ and then continuously increasing St.

Neutral curves giving the critical draw ratio Dr_{c}, the frequency ω_I and the scaled base-state tension f_s, versus the Stanton number St are plotted in Fig. 10.2 for $Bi = 6$, $b = 1$ and four different sensitivities μ of the viscosity with temperature.

A maximum occurs at $St = \mathcal{O}(1)$ and corresponds to the largest critical draw ratio, denoted ‘S’. On the left of this maximum, the critical draw ratio increases with St, indicating a stabilizing effect of the heat transfer. This region will be referred hereafter to the advection-dominated regime, denoted ‘A’. On the right of the maximum, the stability region decreases with increasing the Stanton number. This region will be referred hereafter to the transfer-dominated regime, denoted ‘T’. In the limit of very large St, the critical draw ratio Dr_{c} can even lie below 20.218, which is unusual, at

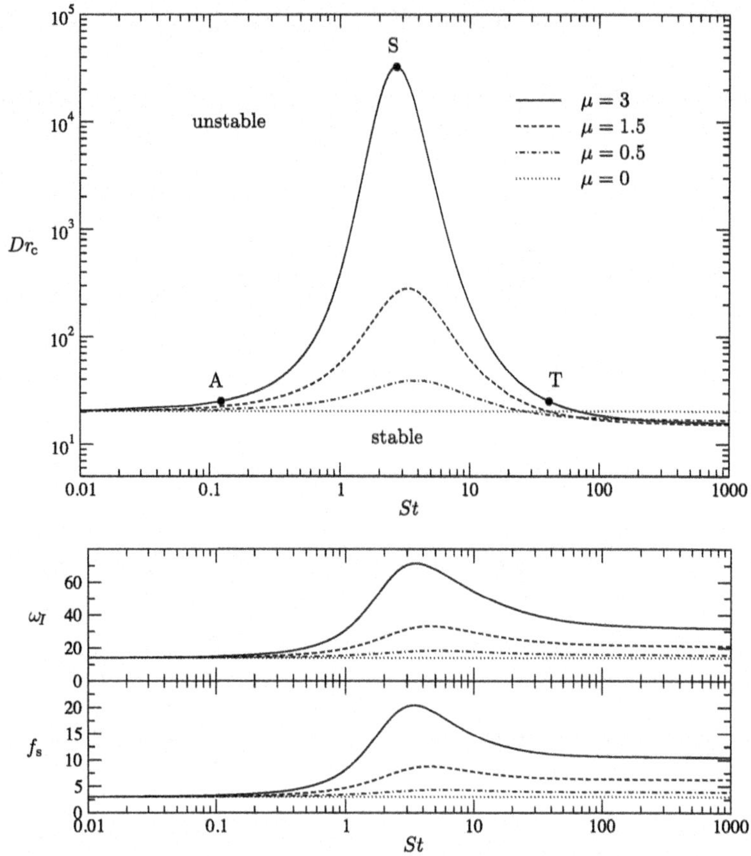

Fig. 10.2 Neutral stability curves for $Bi/6 = b = 1$ and for different values of μ. Filled circles indicate solutions lying in the different regime of cooling: (A) Advection-dominated regime, (T) Transfer-dominated regime, (S) Largest critical draw ratio regime

least for Newtonian film casting: $Dr_c < 20.218$ is indeed possible for film drawing (see Chap. 5) as well as for viscoelastic fluids (see Chap. 8).

As shown in Fig. 10.2, the frequency ω_I and the tension f_s follow the same trend as for Dr_c, except in the limit of very large St where their values are significantly larger than in isothermal conditions. This result means that a film with increasing viscosity along its stretching direction has a smaller period $P = 2\pi/\omega_I$ of draw resonance than for a constant viscosity. Furthermore, except for this limit that will be analyzed in Sect. 10.5, increasing μ always stabilizes the system, as is observed commonly in fiber drawing and film casting.

Next, we compare the model accounting for a parabolic temperature profile across the film with the one-dimensional model assuming a constant temperature profile. The latter being recovered by simply setting $Bi \to 0$ in (10.16). Note it does not mean that the heat transfer is vanished as the Bi number is also incorporate in the

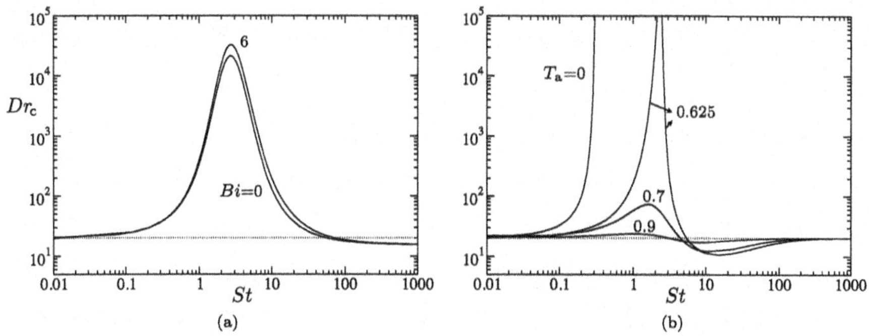

Fig. 10.3 Neutral stability curve for $\mu = 3$. **a** Influence of the Biot number with $T_a = e^{-x}$, i.e. $b = 1$. **b** Influence of constant ambient temperatures T_a with $Bi \to 0$

St number, which is non-zero. The neutral stability curves for $\mu = 3$ are plotted in Fig. 10.3a, which shows that temperature variations across the film are stabilizing. Though this effect is small, it is obviously more pronounced as the Stanton number increases, i.e. in the region of large heat transfer ($St \gtrsim 1$). Of course, temperature variations across the film should vanish in the limit of $St \to \infty$, for which the liquid temperature is everywhere identical to the ambient temperature T_a.

Figure 10.3b also shows the consequences of setting the ambient temperature to a constant. For $T_a = 0$ and $Bi \to 0$, (10.16) reduces to (10.19) as studied by Shah and Pearson (1972a) for fiber drawing. As found by these authors, the critical draw ratio rises abruptly as St increases, still in the region $St \ll 1$, leaving on the right of the neutral curve a large region of unconditional stability.[5]

However, the divergence of the neutral curve for $T_a = 0$ should only be attributed to the concomitant divergence of the viscosity as the fluid temperature approaches T_a, i.e. $\theta \to 0$ (see (10.21)), when the Stanton number is increased. Figure 10.3b then shows that there is a range of finite values for T_a for which the neutral curves can be computed numerically up to large values of St, where they in turn all converge to the isothermal critical draw ratio $Dr_c = 20.218$: in the limit of infinite heat transfer coefficient, and for constant ambient temperature, the system exactly recovers its isothermal state.

We next examine typical solutions for the steady-state and amplitudes of the perturbations. The results are plotted in Fig. 10.4 and correspond to the two different regions of the neutral stability curve for $\mu = 3$ in Fig. 10.2.

In the region $St \ll 1$, the solution ‘A’ (dashed lines) in Fig. 10.4 shows a gentle temperature decrease along the stretching direction, hence the viscosity at take-up is only 12% larger than the inlet viscosity, so that the stretched profiles (u_s, h_s) resemble those for isothermal stretching. Likewise, the corresponding amplitudes of

[5] This result was nevertheless corrected for fiber drawing by Gupta et al. (1996) who showed the system to be always conditionally stable. These authors used instead a heat transfer coefficient that depends on the velocity, but still with a constant ambient temperature.

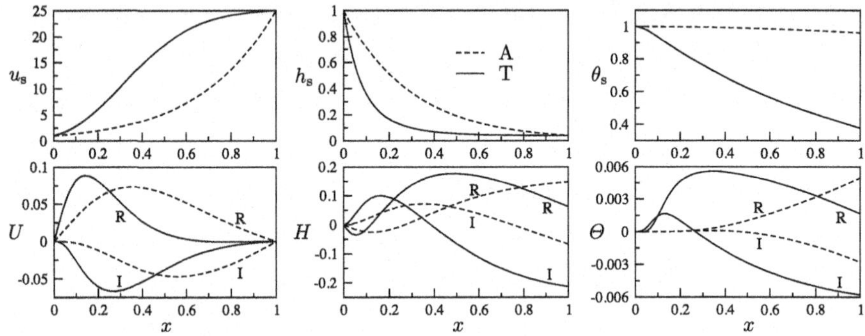

Fig. 10.4 Solutions to the eigenvalue problem (10.25–10.27) at criticality for $Dr_c = 25$ and $\mu = 3$, as identified in Fig. 10.2: (A) $St = 0.123$ (*dashed lines*) and (T) $St = 40.92$ (*solid lines*). The steady state solutions are above and the complex-valued perturbations below, with 'R' their real parts and 'I' their imaginary parts

the perturbations have no significant differences from the isothermal case. On the contrary, in the region of $St \gg 1$, the solutions (solid lines) show a different behavior, which is determined mainly by the strong decrease of the temperature θ_s that makes the viscosity at take-up about 140 times the inlet value. Most of the stretching thus occurs in the region of low viscosity as shown from the solutions for h_s and u_s in Fig. 10.4. The amplitudes of the perturbations are also distributed differently than those for isothermal conditions. The most significant difference is the real part of the velocity perturbation U (thick line), which reaches its maximum in the first-half of the film where the viscosity is the lowest, and decreases rapidly to zero in the second half. Though less obvious, this behavior is also the fate of the thickness perturbation, keeping in mind from (10.23) that the real perturbation is the product of H with the base state h_s, which is minimum in the second half of the film.

Finally, in terms of the cooling process, and based on the definition of the Stanton number (10.17), we can summarize our observations as follows: (i) The region of $St \ll 1$ corresponds to an *advection-dominated* cooling for which an increase of the heat transfer coefficient stabilizes the system with respect to the draw resonance instability. These conditions are the usual situations found in industry for film casting and fiber drawing, where the system is found to be stable for draw ratios as large as $Dr = \mathcal{O}(10^5)$. Typical Stanton numbers are smaller than unity, $St < 1$, and correspond to 'fast' flows, $Pe = \mathcal{O}(100)$, and large film parameter, $\beta = \mathcal{O}(10^{-1})$. Note the heat transfer coefficient in this regime can still be very high, i.e. $Bi = \mathcal{O}(1)$. (ii) The region of $St \gg 1$ corresponds to a *transfer-dominated* cooling, which, in contrast, is destabilized by an increase of the heat transfer coefficient. Practical examples would correspond to 'slow' flows, $Pe = \mathcal{O}(10)$, and small film parameter, $\beta = \mathcal{O}(10^{-3})$, together with a large heat transfer coefficient. Typical processes that meet such conditions are found in the glass film industry where the Stanton number is commonly of $\mathcal{O}(100)$.

10.5 Prescribed Temperature Limit

As mentioned in Sect. 10.2.2, in the limit of $St \to \infty$, the temperature of the film obtains the prescribed temperature of the ambient, i.e. $\theta_s = T_a = e^{-bx}$. If for the sake of simplicity, we further assume that the far-field temperature slightly decreases along the stretching direction, i.e. $b \ll 1$, then the temperature can be written as $\theta_s \approx 1 - bx$, which once substituted in (10.21) yields $\hat{\eta} \approx e^{\mu b x}$. Consequently, the viscosity is a prescribed function of position that only depends on the product μb and the stability problem can then be solved independently of the temperature equation and with μb as the only free parameter.

Neutral stability curves and solutions for this case are plotted in Fig. 10.5. For an exponential increase of the viscosity along the stretching axis, the results show that the critical draw ratio can be significantly below 20.218. However, as μb increases, the neutral curve saturates at $Dr_c \approx 16.5$, while the frequency of the instability increases quasi-linearly. Solution 'c' in Fig. 10.5 corresponds to a film whose viscosity at take-up is $e^{10} \approx 22000$ times the inlet viscosity. Such a large difference is common in the processing of highly viscous films.

Because the viscosity is an explicit function of position, $\hat{\eta}(x)$, there is an analytical solution for the base state,

$$u_s(x) = e^{f_s \int_0^x \frac{dx'}{\hat{\eta}(x')}} \qquad \text{with} \quad f_s = \frac{\ln Dr}{\int_0^1 \frac{dx}{\hat{\eta}(x)}}, \tag{10.29}$$

and consequently the residence time $\tau_L = \int_0^1 (1/u_s)\, dx$ of the fluid from the inlet to the take-up can be straightforwardly calculated. Since $h_s = 1/u_s$, we can observe

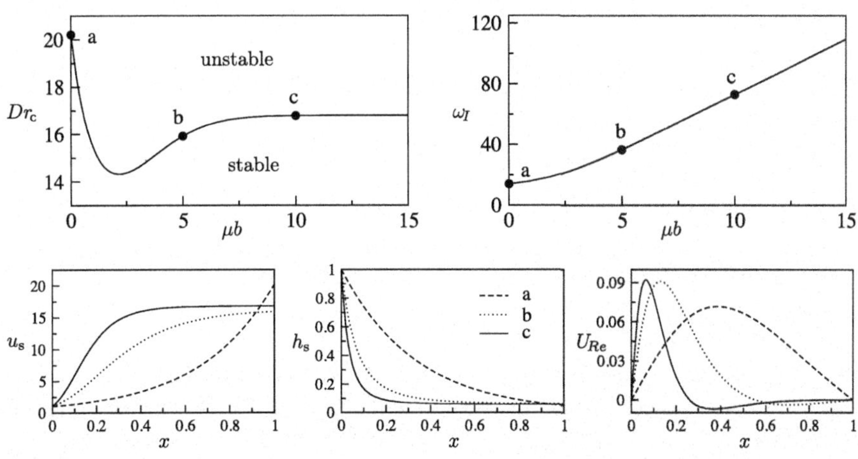

Fig. 10.5 (*Top*) Neutral stability curves for Dr_c and ω_I in the 'prescribed temperature' limit. (*Bottom*) Typical solutions relative to μb

from Fig. 10.5 that the residence time decreases as the viscosity contrast μb increases; the liquid obtains a velocity close to the take-up speed in a shorter distance. For instance, the solution 'a' ($\mu b = 0$) in Fig. 10.5 corresponds to $\tau_L = 0.32$, while solution 'c' ($\mu b = 10$) corresponds to $\tau_L = 0.1$, which yields a factor three between the two residence times. According to *Jung's criterion* (see Sect. 3.3.1), also discussed in the next section, a decrease of the residence time should be stabilizing, and this is the usual effect observed in the advection-dominated region. But as we have found in this study, this criterion is valid only if the period of draw resonance, $P = 2\pi/\omega_I$, remains unchanged as compared to the isothermal conditions. From Fig. 10.5, we rather see the frequency ω_I to be about three times larger for $\mu b = 10$ than for a constant viscosity ($\mu b = 0$). We conjecture here that this strong frequency increase of the draw resonance is related to the fact that the stretching naturally occurs in the region close to the inlet where the viscosity is the lowest. In other words, these results are consistent with the idea that in the large Stanton number limit, a decrease of the *effective stretching length* is accompanied by an increase of the frequency of draw resonance.

10.6 Mechanism of Draw Resonance in the Non-isothermal Case

As compared to the isothermal case for which the domain-averaged unity-throughput wave was shown to be in phase with the perturbation of the tension (see (3.35)), integrating the perturbation Eq. (10.27b) along the domain leads to

$$\frac{\int_0^1 \frac{Q+\Gamma}{\eta_s}\,dx}{\int_0^1 \frac{1}{\eta_s}\,dx} = F, \tag{10.30}$$

where $Q = U + H$ is the perturbation of the flow rate, $\Gamma = -\mu\Theta/\theta$ is the perturbation of the viscosity and $\eta_s = \hat{\eta}(\theta_s)$. This relation demonstrates that contrarily to the isothermal case, it is the weighted mean of $(Q + \Gamma)/\eta_s$ that is in phase with the tension f, and not simply Q. This result thus means that an increase of the viscosity along the film will induce a phase shift between the tension perturbation and the response of the averaged flow rate. This feature turns out to be crucial in understanding the (de)stabilization of draw resonance due to cooling as detailed below.

Based on the mechanism for draw resonance in the isothermal case outlined in Sect. 3.3, we thus propose here a scenario that simultaneously explains the stabilizing mechanism of draw resonance due to cooling, enhanced for increasing μ, and the destabilizing effect of strong viscosity variations along the stretching axis in the prescribed temperature limit.

Like the tension, averaged quantities do not depend on x coordinate, but only depend on time. They can thus be directly related to $f(t)$, especially in terms of

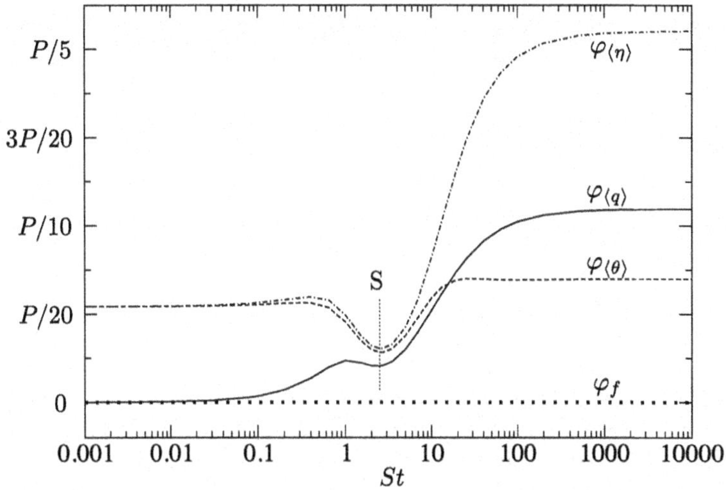

Fig. 10.6 Phase shifts of different averaged quantities relative to the tension and normalized by the period P. The results correspond to the case $\mu = 3$ in Fig. 10.2. 'S' refers to the stabilized regime

the phase shift, based on the principle that the tension leads the mechanism of draw resonance. Consequently, we define $\langle X\rangle(t) = \int_0^1 X(x,t)\,dx$ as the axial-average of quantity X at a given time t. Also, we write $\varphi_{\langle X\rangle}$ for the phase shift of this averaged quantity relative to oscillations of f taken as a reference, i.e. $\varphi_f = 0$, and normalized with the period P of oscillations. The general expression for the phase shift is given by (3.44).

Figure 10.6 shows, as predicted by (10.30), that the axial-averaged flow rate $\langle q\rangle$ is in phase with the tension in the region of $St \ll 1$, and is delayed for increasing Stanton numbers. In the 'stabilized regime' for $St = \mathcal{O}(1)$, i.e. at the transition S on Fig. 10.2, the averaged temperature $\langle\theta\rangle$ is nearly in phase with $\langle q\rangle$. The explanation for the stabilizing effect of cooling in this region of the parameter space is thus as follows: a perturbative increase of the tension creates, after a small delay $\varphi_{\langle q\rangle}$, an increase of the average flow rate, which implies that in average, the liquid spends a little less time in the system and hence has a little less time to cool down. The averaged temperature $\langle\theta\rangle$ thus increases in response to an increase of the averaged flow rate. At the same time, $\varphi_{\langle\eta\rangle}$ and $\varphi_{\langle\theta\rangle}$ being nearly in phase in the stabilized regime, the averaged viscosity $\varphi_{\langle\eta\rangle}$ decreases accordingly, which in turn 'relaxes' the initial positive perturbation of the tension. This stabilization mechanism due to cooling is illustrated below for a positive perturbation of the tension,

$$\text{Perturbation} \Rightarrow f\nearrow \Rightarrow \langle q\rangle\nearrow \Rightarrow \langle\theta\rangle\nearrow \Rightarrow \langle\eta\rangle\searrow \Rightarrow f\searrow \Rightarrow \text{Damping}\,. \tag{10.31}$$

The same reasoning can be done for a negative perturbation, which is thus followed by an increase of the averaged viscosity, hence again a counter-action to the initial perturbation. This mechanism shows how the cooling process is stabilizing, or in other words why the critical draw ratio for draw resonance is much larger in the region of $St = \mathcal{O}(1)$, as compared to isothermal conditions.

It is now easy to see that the above mechanism is most efficient in the region where the energy advected by the flow is comparable to the heat transferred to the surroundings, i.e. at the transition 'S' as labeled in Fig. 10.6. Indeed, in both limits of $St \to 0$ and $St \to \infty$, the temperature perturbation vanishes and the system behaves as if it was isothermal (see discussion in Sect. 10.2.2). The fact that the amplitude of the temperature perturbation vanishes in those two limits can be seen from (10.27c) where

$$\text{if} \begin{cases} St \to 0 \\ St \to \infty \end{cases} \Rightarrow \quad \Theta \to 0 \quad \text{since} \begin{cases} \Theta(0) = 0 \\ \theta_s \to T_\text{a} \end{cases} . \tag{10.32}$$

Note finally that for 'advection-dominated' cooling ($St \ll 1$), the averaged viscosity and temperature are exactly in phase, as for isothermal conditions. On the contrary for 'transfer-dominated' cooling ($St \gg 1$), the averaged viscosity is delayed relative to the averaged temperature and the phase shift with the tension is approaching a quarter of a period, which makes ineffective the stabilizing response of the averaged viscosity with tension perturbations.

10.7 Stability Criteria in Non-isothermal Conditions

In this section, we want to illustrate how *Jung's criterion* is modified in non-isothermal conditions. To this aim, we have reported in Fig. 10.7 the different characteristic times (3.29,3.41,3.43) defined in Sect. 3.3.

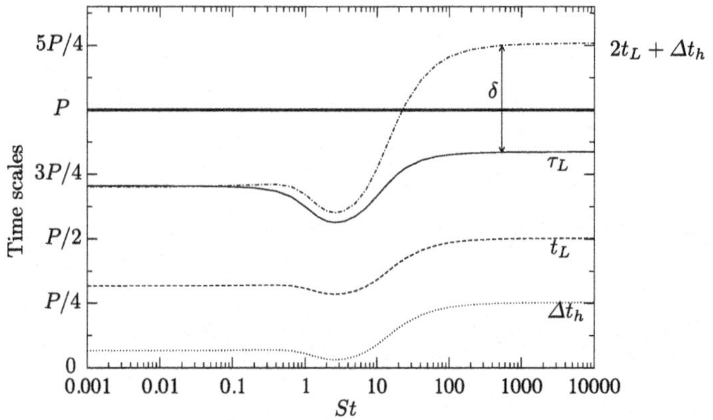

Fig. 10.7 Characteristic times of the system normalized with the period P of draw resonance: τ_L the residence time, t_L the unity-throughput time, Δt_h the tension/thickness phase shift, and δ the viscous delay. The results correspond to the case $\mu = 3$ in Fig. 10.2

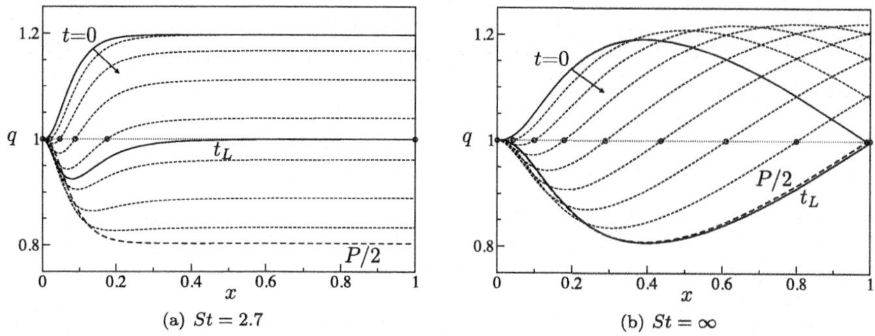

Fig. 10.8 Time evolution of the flow rate q at criticality, for two different Stanton numbers St and for $Bi/6 = b = 1$ and $\mu = 3$. Same legend as for Fig. 3.7 with **a** $Dr_c = 32372$, $t_L = 0.026$, $P = 0.092$ and **b** $Dr_c = 15.5$, $t_L = 0.1$, $P = 0.198$

We observe that the criterion (3.36) established for isothermal conditions is modified in non-isothermal conditions as follows,

$$2t_L + \Delta t_h < \tau_L + \delta \ , \tag{10.33}$$

where δ accounts for the delay induced by viscosity variations along the film. Note that δ is negligible in the 'advection-dominated' regime so that *Jung's criterion* still applies in this region. On the contrary, δ becomes large (of the order of t_L) in the 'transfer-dominated' regime, indicating a dramatic effect of the viscosity variations on the stability properties. We also note by comparing Figs. 10.6 and 10.7 that δ and $\varphi_{\langle q \rangle}$ are closely correlated, in fact $\delta \approx 4\varphi_{\langle q \rangle}$. In order to illustrate this correlation, we show in Fig. 10.8 the time evolution of the flow rate for two different Stanton numbers corresponding to the most stable regime, $St = \mathcal{O}(1)$, and to the 'prescribed-temperature' regime, $St \to \infty$.

We notice that the flow rate in the first third of the domain in Fig. 10.8a behaves similarly to the isothermal case in the entire domain (see Fig. 3.7)—in the other two thirds, the flow rate is spatially constant and varies only with time. The critical draw ratio is extremely high in this case and the velocity in the constant-flow-rate region is so high that the residence time is negligible as compared to the time spent by the fluid in the first third of the domain. Consequently, the 'effective stretching' length is shortened by a factor 3, which is in favor of stability. We also observe that the unity-throughput wave accelerates strongly along x in Fig. 10.8a for the reason given above, while it decelerates in the limit $St \to \infty$ (Fig. 10.8b) due to the increase of the viscosity along the stretching direction.[6] This effect is precisely measured by the

[6] Note the time evolution corresponding to the advection-dominated regime is qualitatively similar to the time evolution in isothermal conditions ($St = 0$) already illustrated in Fig. 3.7.

time shift δ, however this parameter cannot be predicted from the linear theory. On the contrary, the *Kim's criterion* (3.34) remains valid in the non-isothermal case as it can easily be verified in Fig. 10.7 where we can observe a constant shift equivalent to $P/4$ between Δt_h and t_L for the entire range of St.

10.8 Radiative Effect

To some extent, this 'heat transfer-dominated' regime can be associated with a cooling dominated by radiative heat transfer rather than by convective heat transfer, as is the case for manufacturing glass sheets. This statement is supported by the work of Willien et al. (1988) who found in fiber drawing that radiative heat transfer can be destabilizing, which is opposite to the response for the case where convective heat transfer is dominant. For the sake of simplicity, we only consider the one-dimensional equation, that reads for radiative heat transfer as follows

$$\rho c_p h(\partial_t \theta + u \partial_x \theta) = -2\varepsilon_T \sigma_{SB} \left(\theta^4 - T_a^4\right), \tag{10.34}$$

where ε_T is the total emissivity of the molten liquid and σ_{SB} is the Stefan-Boltzmann constant. In dimensionless form, using the same scaling rules as in Sect. 10.2.2, (10.34) becomes

$$\partial_t \theta + u \partial_x \theta = -St \, \frac{\theta^4 - T_a^4}{4h}, \tag{10.35}$$

where the Stanton number has the same definition than in (10.17) provided the heat transfer coefficient α is substituted by $4\varepsilon_T \sigma_{SB} T_0^3$. We then note that the radiative heat transfer can be linearize for θ around T_a, such as it gives a term that has almost the same functional than the convective heat transfer, namely

$$\partial_t \theta + u \partial_x \theta = -St \, T_a^3 \frac{\theta - T_a}{h}. \tag{10.36}$$

Nevertheless, it is only for constant T_a that the correspondence would be perfect between radiative and convective heat transfer, even though this hypothesis is definitely unrealistic. Instead having θ close to T_a is a reasonable assumption that makes (10.36) a good candidate for future investigations. Note however that radiative heat transfer will be treated in more general terms in the next chapter for fiber drawing.

Chapter 11
Non-isothermal Fiber Drawing

Abstract In this chapter, we investigate the draw resonance of non-isothermal fiber drawing. To that aim we propose a non-isothermal one-dimensional model encompassing inertia, gravity and surface tension as well as convective and radiative heat transfer. The practical scaling introduced in Chap. 9 is adapted to handle this multi-parametric problem by isolating the control parameters relevant to practical applications, namely the inlet temperature and the fiber aspect ratio, in a real set-up. A linear stability analysis is then performed in order to produce neutral stability curves by numerical continuation in a way which permits to explore the full parameter space. The influence of the viscosity law, the ambient temperature decrease and the varying heat transfer coefficient on the stability are first determined using the pure viscous model. The stability behavior for the full model is then investigated and with an emphasis on the effect of the aspect ratio, the inlet temperature and the stiffness of the decreasing ambient temperature along the fiber. Each of the parameters used in this chapter has been chosen in a range relevant to glass fiber manufacturing.

11.1 Non-isothermal Model

In this chapter, we consider the glass fiber drawing to address the non-isothermal effect, which corresponds to a configuration very close to application. The geometry given in Fig. 2.1 is adapted to non-isothermal case in Fig. 11.1, where the feeding manifold has been added.

The filament is fed at its top by molten glass at temperature θ_0 through a cylindrical nozzle of radius R_0 and height ℓ at velocity $u(0, t) = u_0(\theta_0)$ as sketched in Fig. 11.1. The exit point of the nozzle is referred to as the inlet. The filament of cross-section $a(x, t)$ is still assumed axisymmetric. With molten glass being the only material considered in this chapter, we assume that the liquid is Newtonian and that its temperature-dependent dynamic viscosity $\eta(\theta)$ is given by the Vogel-Fulcher-Tamman (VFT) law, which is widely used in the field of glass processing and given by Scherer (1992),

© Springer Nature Switzerland AG 2025

M. Bechert and B. Scheid, *Draw Resonance*, Fluid Mechanics and Its Applications 141,
https://doi.org/10.1007/978-3-031-85568-9_11

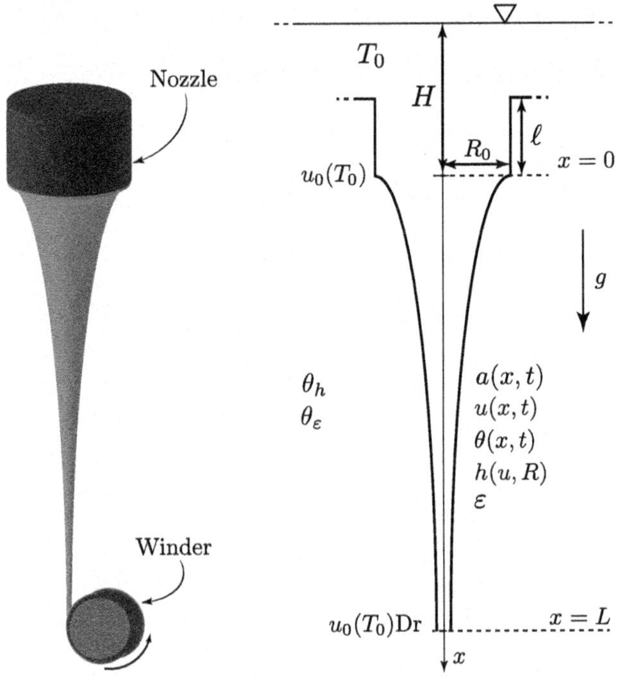

Fig. 11.1 Left: Sketch of the fiber drawing process. Right: One-dimensional non-isothermal fiber model (see text for details)

$$\log \eta(\theta) = \log \eta_\infty + \frac{B}{\theta - \theta_c}, \tag{11.1}$$

where $\theta = \theta(x, t)$ denotes the cross-section-averaged temperature of the fiber and η_∞, B and θ_c are three constants which depend on the glass composition. In all generality this framework can be applied to any kind of Newtonian molten materials. The evolution of the temperature field is governed by both convective and radiative heat transfer. The heat transfer coefficient h is not constant. Indeed it appears that it depends explicitly on the velocity and on the cross-section of the fiber, following previous studies (Beyreuther and Brünig, 2007; Kase and Matsuo, 1965). As we consider the stability of a single fiber in the present chapter, which excludes chaotic air motion occurring in fibers bundle, the air flow induced by the process is expected to be laminar. In this regime, the most commonly used model in the literature (Shah and Pearson, 1972b; Gupta et al., 1996) that provides an explicit relation $h = h(a, u)$ has been proposed by Kase and Matsuo (1965). These authors measured experimentally the heat transfer coefficient of a heated wire in a air flow and obtained for the particular case of a flow parallel to the fiber a relation of the following form:

$$h(a,u) = \frac{0.42\pi^{1/3}k_a}{\nu_a^{1/3}}\left(\frac{u}{a}\right)^{\frac{1}{3}}, \tag{11.2}$$

where k_a and ν are respectively the thermal conductivity and kinematic viscosity of air. As mentioned above, in many practical applications, fiber drawing can involve a bundle of several thousand of fibers. In such cases, the air flow can be chaotic and an explicit modeling for the heat transfer coefficient can be challenging. Molten glass is here considered as an opaque medium and internal radiation is neglected (Chouffart et al., 2016). Therefore radiative heat transfer is modeled through a constant effective emissivity of glass fiber surface ε.[1] The convective and radiative heat transfer are parametrized by a reference temperature, θ_h and θ_ε, respectively. Indeed in a manufacturing process, thousands of fibers are produced simultaneously and it is not possible to define the ambient temperature as the temperature far from the fiber. Therefore it is necessary to define the external temperature through process-dependent approximations leading to θ_h and θ_ε. These estimations correspond to experimental or numerical measurements obtained in the region where each of the contributions is dominant. As radiative and convective exchanges are dominant close to the nozzle and along the fiber respectively, we impose two different reference temperatures, as considered by Chouffart et al. (2016).

By adopting the same scaling as in Sect. 9.1, the free-fall velocity $\sqrt{gL}$ is used as reference velocity. This choice permits here to set the velocity scale directly from the chosen cast length L. Likewise, all variables are non-dimensionalized according to the following transformations :

$$x \to Lx\,,\quad a \to R_0^2 a\,,\quad t \to \sqrt{\frac{L}{g}}\,t\,,\quad u \to \sqrt{gL}\,u\,,\quad \eta \to \eta_0\eta\,, \tag{11.3a}$$

$$h \to h_0\,h\,,\quad \varepsilon \to \varepsilon_0\,\varepsilon\,,\quad (\theta,\theta_h,\theta_\varepsilon) \to \theta_c + \Delta\theta\,(\theta,\theta_h,\theta_\varepsilon)\,, \tag{11.3b}$$

where $\eta_0 = \eta(\theta_0)$, $h_0 = 0.42\pi^{1/3}k_a(gL)^{1/6}(\nu_a R_0^2)^{-1/3}$, ε_0 is a typical average emissivity along the fiber in the vicinity of the nozzle (Chouffart et al., 2016) and $\Delta\theta = \theta_0 - \theta_c$. The resulting dimensionless parameters are defined by the following expressions (repeating those already obtained in Sect. 9.1 for completeness):

$$Q = \frac{u_0}{\sqrt{gL}}\,,\quad F = \frac{\rho\sqrt{gL^3}}{\eta_0}\,,\quad S \equiv \left(\frac{\ell_c}{\sqrt{2R_0L}}\right)^2 = \frac{\gamma}{2\rho g R_0 L}\,, \tag{11.4a}$$

$$St = \frac{h_0}{\rho c_p R_0}\sqrt{\frac{L}{g}} \quad\text{and}\quad St_r = \frac{\varepsilon_0\sigma\Delta\theta^3}{\rho c_p R_0}\sqrt{\frac{L}{g}}\,, \tag{11.4b}$$

[1] It appears that the contribution of radiative effects is only important in the first few millimeters below the inlet (Huynh and Tanner, 1983). In this region the variation of the effective emissivity due to the temperature decrease compensates the variation due to the radius decrease (Chouffart, 2018), which justifies the use of a constant ε.

Table 11.1 Dimensional and dimensionless parameters used for the non-isothermal fiber drawing problem. Parameters that are varied are in bold

Dimensional	Value	Dimensionless	Value
$\boldsymbol{L}$ **(m)**	$\mathbf{6 \times 10^{-2} - 1.8}$	$\boldsymbol{\Lambda}$	$\mathbf{100 - 4000}$
$\boldsymbol{\theta_0}$ **(K)**	$\mathbf{1543.15 - 1603.15}$	$\boldsymbol{F^*}$	$\mathbf{10^{-3} - 3 \times 10^{-3}}$
$\boldsymbol{h_0}$ **(W/m²/K)**	$\mathbf{2.805 \times 10^{-3} - 2805}$	**St***	$\mathbf{10^{-8} - 10^{-2}}$
ε_0 (–)	0.4	St_r^*	5.8×10^{-5}
γ (N/m)	0.39	S^*	22.4
B (°C)	5262.4	B^*	5.86
H (m)	0.35	θ_c^*	0.45
R_0 (m)	6×10^{-4}	q	10.4
ℓ (m)	4.2×10^{-3}	k	1–30
η_∞ (Pa.s)	$10^{-4.075}$	θ_{h_∞}	0.22
θ_c (°C)	402.3	$\theta_{\varepsilon_\infty}$	–0.22
ρ (kg/m³)	2470		
c_p (J/kg/K)	1480		

where Q, F, S, St and St_r are respectively the dimensionless inlet velocity, fluidity, surface tension parameter and the convective and radiative Stanton numbers. Parameter S compares the capillary length $\ell_c = \sqrt{\gamma/(\rho g)}$ to the geometric mean of the fiber dimensions while the Stanton numbers compare the convective and, respectively, the radiative heat transfer, to the energy advected by the flow; σ is the Stefan-Boltzmann constant and ρ, γ and c_p denote density, surface tension and specific heat capacity of the molten glass, respectively (see Table 11.1). Due to a weak dependence on temperature, each of these three last physical parameters are still assumed to be constant. As it could be difficult to properly define the cast length L in industrial applications because of the unknown location of the glass transition point, it is more convenient from now on to consider it as a control parameter of the problem. Hence by introducing the aspect ratio of the fiber $\Lambda = L/R_0$, we propose a new set of dimensionless parameters which do not involve the drawing length L anymore :

$$Q^* \equiv \sqrt{\Lambda}\, Q = \frac{u_0}{\sqrt{g R_0}}, \quad F^* \equiv \frac{F}{\Lambda^{\frac{3}{2}}} = \frac{\rho\sqrt{g R_0^3}}{\eta_0}, \quad S^* \equiv \Lambda\, S = \left(\frac{\ell_c}{\sqrt{2R_0^2}}\right)^2 = \frac{\gamma}{2\rho g R_0^2}, \tag{11.5a}$$

$$St^* \equiv \frac{St}{\sqrt{\Lambda}} = \frac{h_0}{\rho c_p \sqrt{g R_0}} \quad \text{and} \quad St_r^* \equiv \frac{St_r}{\sqrt{\Lambda}} = \frac{\varepsilon_0 \sigma \Delta\theta^3}{\rho c_p \sqrt{g R_0}}. \tag{11.5b}$$

Consequently each of these numbers permit to control individually an experimentally accessible quantity. More precisely Q^*, F^*, S^*, St^* and St_r^* respectively control $u_0(\theta_0)$, $\eta_0(\theta_0)$, γ, h_0 and ε_0. The gauge of the heat transfer coefficient, using (11.2) at the inlet, becomes $h_0 = 0.42\pi^{1/3} k_a \nu_a^{-1/3} R_0^{-1/2} (g\Lambda)^{1/6}$. In this configuration, it appears that the temperature at the inlet θ_0 is a control parameter of the problem as it fixes the value of u_0 and η_0, thus Q^* and F^*. As mentioned above S^* will be kept constant in the following. Unless specified otherwise, all quantities in the rest of the chapter are dimensionless.

Here we assume, as in practical cases, that the flow of molten glass through the cylindrical nozzle is a Poiseuille flow driven by an hydrostatic pressure difference, as illustrated in Fig. 11.1. Therefore the velocity at the inlet is not a control parameter and is rather given, using the Hagen-Poiseuille law, by $u_0(\theta_0) = \rho g H R_0^2/(8\ell\eta_0(\theta_0))$ where H is the height of molten glass above the nozzle. Hence the dimensionless inlet velocity becomes $Q^* = (H/8\ell)\, F^*$. More generally it is possible to obtain such a linear relationship between Q^* and F^* for any geometry of the nozzle,

$$Q^* = qF^* \,, \tag{11.6}$$

where q is a shape factor. Considering the characteristic dimensions of nozzle geometries and molten glass height which could be found in industrial applications, we set $q = H/8\ell = 10.4$ for the shape factor. Remarkably, the dimensionless inlet velocity Q^* is not an independent parameter when the molten liquid is injected by a fixed hydrostatic pressure difference.

It has been shown in Chap. 10 that a non-monotonic behavior of the neutral stability occurs when increasing the Stanton number from the advection-dominated to the heat transfer-dominated regime, which denote the low and high Stanton number regimes respectively. The latter occurs for film casting when $St_{\text{film}} = 2\Lambda\, St^*/Q^* \gtrsim 1$, i.e. when $St^* \gtrsim Q^*/\Lambda$ in the present scaling.[2] In this regime, cooling may be destabilizing when the Stanton number is increased. The same phenomenon has also been evidenced for radiative heat transfer (Geyling and Homsy, 1980; Willien et al., 1988).

In practical applications, the ambient temperature in the vicinity of the fiber is non-homogeneous. Similarly to the film casting in Sect. 10.2.4, it is assumed that the ambient temperature θ_h considered for convective heat transfer follows an exponential decay along the fiber, i.e.

$$\theta_h = \theta_{h_\infty} + (1 - \theta_{h_\infty})\,\mathrm{e}^{-kx} \,, \tag{11.7}$$

where k is a stiffness parameter and θ_{h_∞} is the dimensionless far field temperature associated to convective heat transfer. As an additional justification of (11.7), since the fiber temperature should take the ambient one in the infinite Stanton number limit, this ambient temperature could not reasonably be constant, unless leading to irregular solutions arising due to the incompatibility with imposing the boundary condition $\theta(0, t) = 1$. Finally, we impose a constant radiative temperature $\theta_\varepsilon = \theta_{\varepsilon_\infty}$ with $\theta_{\varepsilon_\infty}$ the dimensionless far field temperature.

---

[2] To avoid confusion, we use subscript 'film' for the Stanton number defined by (10.17).

We further hypothesize that the temperature profile is assumed uniform along the radial coordinate of the fiber. Indeed, it has been shown in Fig. 10.3b for film casting that the influence of the temperature profile across the film thickness on the stability of the system is weak. Additionally, the second-order heat diffusion effect in the longitudinal direction is also neglected for the same reason, as shown for fiber drawing by (Gupta et al., 1996). Under these assumptions, the governing system of equations for the non-isothermal fiber drawing can be rewritten by combining, using the scaling presented above, the cross-section averaged continuity Eq. (2.16) and momentum Eq. (9.5), with the time-dependent version of the leading-order energy equation written in Taroni et al. (2013):

$$\partial_t a + \partial_x (au) = 0 , \tag{11.8a}$$

$$F^* \Lambda^{\frac{3}{2}} a \left(\partial_t u + u \partial_x u - 1 - \frac{S^*}{\Lambda} \frac{\partial_x a}{a^{\frac{3}{2}}} \right) = \partial_x (3 \eta a \partial_x u) , \tag{11.8b}$$

$$\partial_t \theta + u \partial_x \theta = -2 \sqrt{\frac{\Lambda}{a}} \left[St^* \, h(a,u)(\theta - \theta_h) + St_r^* \left(\left(\theta_c^* + \theta\right)^4 - \left(\theta_c^* + \theta_\varepsilon\right)^4 \right) \right] , \tag{11.8c}$$

with $\theta_c^* = \frac{\theta_c}{\Delta\theta}$. This system is closed through the dimensionless VFT law which couples the momentum Eq. (11.8b) and the energy Eq. (11.8c) :

$$\eta(\theta) = 10^{B^* \left(\frac{1}{\theta} - 1\right)} , \tag{11.9}$$

where $B^* = \frac{B}{\Delta\theta}$, as well as with the dimensionless Kase and Matsuo law :

$$h(a,u) = \left(\frac{u}{a}\right)^{\frac{1}{3}} . \tag{11.10}$$

The associated boundary conditions are given at any time t by:

$$a(0,t) = 1 , \quad u(0,t) = \frac{q F^*}{\sqrt{\Lambda}} , \quad u(1,t) = \frac{q F^*}{\sqrt{\Lambda}} Dr , \quad \theta(0,t) = 1 . \tag{11.11}$$

In the case of $St^* \to 0$ and $St_r^* \to 0$ we recover the isothermal fiber drawing problem, but with additional flexibility of independently varying the dimensionless drawing length Λ.

Parameter values used in this chapter are displayed in Table 11.1 and have been chosen according to practical applications involving molten glass (Chouffart et al., 2016). Parameters St_r^*, B^* and θ_c^* have small variations regarding to the range of inlet temperature θ_0 considered. For the sake of simplicity, we impose their value to the one corresponding to $\theta_0 = 1573.15$ K as these variations have a minor influence on the stability curves (data not shown). The ranges of values for the fiber length L and heat transfer coefficient h_0 proposed in this chapter are broader than the ones found in industrial applications in order to widely explore the influence of these control parameters.

11.2 Stability Analysis

We study the stability of the system of Eq. (11.8) following the same procedure as before (see Sect. 3.2). Even though it is not strictly necessary, we also expand for convenience the VFT Eq. (11.9) in the same way. We thus introduce the following expansions :

$$a(x,t) = a_s(x)(1 + A(x)\mathrm{e}^{\omega t}) \,, \tag{11.12a}$$
$$u(x,t) = u_s(x)(1 + U(x)\mathrm{e}^{\omega t}) \,, \tag{11.12b}$$
$$\theta(x,t) = \theta_s(x)(1 + \Theta(x)\mathrm{e}^{\omega t}) \,, \tag{11.12c}$$
$$\eta(x,t) = \eta_s(x)(1 + \Gamma(x)\mathrm{e}^{\omega t}) \,, \tag{11.12d}$$

where the suscript s identifies the steady state, A, U, Θ and Γ represent the complex spatial perturbations, and $\omega = \omega_R + \mathrm{i}\,\omega_I$ is the complex eigenvalue. The complex conjugate terms are omitted for the sake of readability. Within this framework one obtains the following base state equations:

$$a_s' = -\frac{a_s u_s'}{u_s} \,, \tag{11.13a}$$

$$u_s'' = -\frac{a_s' u_s'}{a_s} + \frac{F^* \Lambda^{\frac{3}{2}}}{3\eta_s}\left(u_s' u_s - 1 - \frac{S^*}{\Lambda}\frac{a_s'}{a_s^{3/2}}\right) - \frac{\eta_s' u_s'}{\eta_s} \,, \tag{11.13b}$$

$$\theta_s' = -\frac{2}{u_s}\sqrt{\frac{\Lambda}{a_s}}\left[St^*\,h(a_s,u_s)\,(\theta_s - \theta_h) + St_r^*\,\left(\left(\theta_c^* + \theta_s\right)^4 - \left(\theta_c^* + \theta_\varepsilon\right)^4\right)\right] \,, \tag{11.13c}$$

$$\eta_s = 10^{B^*\left(\frac{1}{\theta_s} - 1\right)} \,, \tag{11.13d}$$

where the prime corresponds to the derivative with respect to x. By introducing the expansions (11.12) into (11.8) and (11.9), and by linearizing the resulting equations for small perturbations, one obtains the following perturbation equations:

$$A' = -\left(\frac{\omega}{u_s} A + U'\right) \,, \tag{11.14a}$$

$$U'' = \frac{\omega u_s'}{u_s^2} A - \frac{u_s'}{u_s}\Gamma' - \frac{\eta_s'}{\eta_s} U' + \frac{F^* \Lambda^{\frac{3}{2}}}{3\eta_s}\left[\left(\frac{1}{u_s} + \omega + u_s'\right) U + u_s U'\right.$$

$$+\frac{S^*}{\Lambda\sqrt{a_s}u_s^2}\left\{\left(\omega-\frac{u_s'}{2}\right)A-u_s'(U+\Gamma)+u_sU'\right\}+\left(\frac{1}{u_s}-u_s'\right)\Gamma\Bigg] , \tag{11.14b}$$

$$\Theta' = -\frac{\omega}{u_s}\Theta - \frac{2\,St^*}{u_s\theta_s}\sqrt{\frac{\Lambda}{a_s}}$$
$$\times\left\{h(a_s,u_s)\theta_h\Theta-(\theta_s-\theta_h)\left[h(a_s,u_s)\left(U+\frac{A}{2}\right)-\left(a_s\partial_{a_s}h(a_s,u_s)A+u_s\partial_{u_s}h(a_s,u_s)U\right)\right]\right\}$$
$$-\frac{2\,St_r^*}{u_s\theta_s}\sqrt{\frac{\Lambda}{a_s}}\left\{4\,\theta_s\left(\theta_c^*+\theta_s\right)^3\Theta-\left(\left(\theta_c^*+\theta_s\right)^4-\left(\theta_c^*+\theta_\varepsilon\right)^4\right)\left(U+\frac{A}{2}+\Theta\right)\right\} , \tag{11.14c}$$

$$\Gamma = -\log(10)B^*\frac{\Theta}{\theta_s} . \tag{11.14d}$$

The steady state satisfies the boundary conditions (11.11) while the perturbations have to vanish where the variables are fixed according to these boundary conditions, i.e.

$$A(0) = U(0) = \Theta(0) = 0 \quad \text{and} \quad U(1) = 0 . \tag{11.15}$$

With the eigenvalue problem (11.14) being linear and homogeneous, the amplitude of the solutions is arbitrary. We thus choose to fix the perturbation of the fiber cross-section A at the outlet to unity, i.e. $A(1) = 1$. This choice has no influence on the eigenvalue spectrum and thus on the calculated stability.

The stability of the system regarding thermal effects is studied by solving the system of ODEs (11.13) and (11.14). For very high values of the critical draw ratio, the stationary cross-section a_s decreases down to extremely small values that can be problematic for the numerical resolution of some terms in (11.13) and (11.14) where it appears at the denominator. A simple way to circumvent this problem is to make use of the mass conservation equation to substitute in these terms the stationary cross-section variable by

$$a_s = \frac{Q^*}{u_s\sqrt{\Lambda}} . \tag{11.16}$$

Following the same procedure as before, we consider the first mode of the instability which corresponds to the eigenvalue with the highest real part. The value of the critical draw ratio Dr_c is deduced by tracking the solution for which $\omega_R = 0$. The neutral stability curves giving the evolution of Dr_c as a function of the Stanton number St^* is obtained by continuation as described in Sect. 3.2.2. Likewise, the pure viscous isothermal solution, i.e., the solution derived for $F^* = St^* = 0$, is used as starting point (see Appendix B).

11.3 Comparison with Film Casting in the Pure Viscous Case ($F^* = 0$) with $h = c^{st}$

Regarding the governing equations, film casting and fiber drawing only differ in the pure viscous case by the right-hand side of the energy equation. Converting (10.19) for film casting to our current set of dimensionless parameters, one obtains

$$\partial_t R_{\text{film}} + \partial_x (R_{\text{film}} u) = 0 \,, \tag{11.17a}$$

$$\partial_x (\eta R_{\text{film}} \partial_x u) = 0 \,, \tag{11.17b}$$

$$\partial_t \theta + u \partial_x \theta = -2 \frac{\sqrt{\Lambda_{\text{film}}}}{R_{\text{film}}} St^* (\theta - \theta_h) \,, \tag{11.17c}$$

where $\Lambda_{\text{film}} = L_{\text{film}}/R_{0,\text{film}}$ is the film aspect ratio with L_{film} the length of the film and $R_{0,\text{film}}$ the film thickness at the inlet. Comparing (11.17c) with (11.8c) for a constant heat transfer coefficient and $St_r^* = 0$, it shows that the right-hand-side is inversely proportional to the film thickness for film casting while it is inversely proportional to $\sqrt{a}$ for fiber drawing. The corresponding neutral stability curves are compared in Fig. 11.2.

Both cases are computed with the Arrhenius law given by (10.21), i.e. $\eta = e^{\mu(\frac{1}{\theta}-1)}$, with $\mu = 3$. The ambient temperature is given by $\theta_h = e^{-x}$ which corresponds to the particular case $\theta_{h_\infty} = 0$ and $k = 1$. We note that the neutral stability curves have qualitatively the same non-monotonic behavior, namely that the heat transfer has a stabilizing effect until the critical draw ratio reaches a maximum and

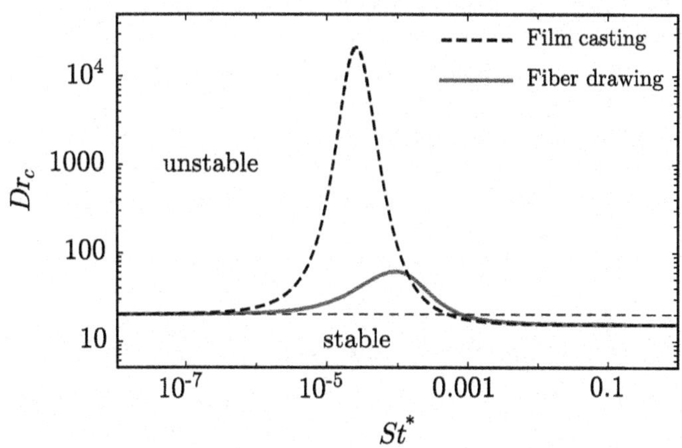

Fig. 11.2 Critical draw ratio as a function of the Stanton number St^* for film casting (*black dashed line*) and fiber drawing (*solid line*) for convective heat transfer only ($St_r^* = 0$) and constant heat transfer coefficient. Each curve is computed for an aspect ratio $\Lambda = \Lambda_{\text{film}} = 10^3$ and an inlet temperature $\theta_0 = 1573.15$ K

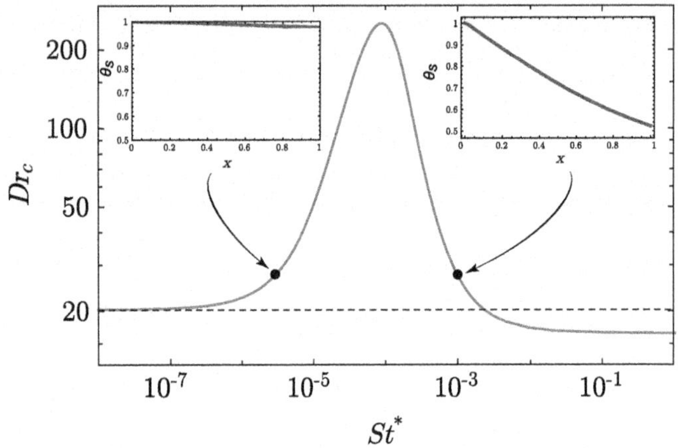

Fig. 11.3 Critical draw ratio as a function of the Stanton number St^* for the pure viscous model with convective heat transfer only ($St_r^* = 0$) and a VFT viscosity law. The curve is computed for a constant heat transfer coefficient, an aspect ratio $\Lambda = 10^3$, an inlet temperature $\theta_0 = 1573.15$ K, $\theta_{h_\infty} = 0.22$ and $k = 1$. The inset represents the steady temperature $\theta_s(x)$ for $St^* = 2.5 \times 10^{-6}$ (*solid line*) and 10^{-3} (*dashed line*) and characterizes the *advection-dominated* and *heat transfer-dominated* regimes respectively

then destabilizes the system as the Stanton number increases. However this maximum Dr_c is two orders of magnitude higher for film casting than for fiber drawing. Additionally, the Stanton number corresponding to the transition from *advection-dominated* to *heat transfer-dominated* regime ($\text{St}^* \sim Q^*/\Lambda$) is also higher for fiber drawing. The critical draw ratio converges towards the same value below 20.218 in the large Stanton number limit in both cases, reflecting the scenario of a prescribed, exponentially decreasing temperature (see Sect. 10.5). All the computing parameters being the same in each case, it is remarkable that such a small variation of the energy equation leads to a strong variation of the critical draw ratio. It also shown in Fig. 11.3 that the change to the VFT viscosity law (compare to the Arrhenius law in Fig 11.2) does not modify the order of magnitude of the maximum Dr_c for fiber drawing.

11.4 Purely Viscous Non-isothermal Model ($F^* = 0$)

In this section we relax some hypotheses of the pure viscous model to understand the spectacular stabilization observed in industrial applications, allowing to draw fiber up to $Dr = \mathcal{O}(10^4)$. We first look at the influence on the stability of the ambient temperature through the stiffness parameter k and second to the heat transfer coefficient. We again neglect inertia, gravity, and surface tension ($F^* = 0$), all addressed in Sect. 11.5, and exclude radiative heat transfer ($St_r^* = 0$), which will be addressed in Sect. 11.5.4. The boundary conditions for the velocity in (11.11) are rewritten in terms of Q^*,

$$u(0,t) = \frac{Q^*}{\sqrt{\Lambda}},\ u(1,t) = \frac{Q^*}{\sqrt{\Lambda}} Dr\ , \tag{11.18}$$

where Q^* has been fixed to $Q^*(\theta_0 = 1573.15\,\mathrm{K}) \approx 1.93 \times 10^{-2}$.

11.4.1 *Influence of the Ambient Temperature*

The choice of the spatial-dependent ambient temperature appears to have a strong influence on the stability of the system. As shown in Fig. 11.4a, the value of the maximum critical draw ratio rises significantly by increasing the value of the stiffness parameter k.

At some point, this maximum becomes even higher than the limit of the domain covered by the numerical computations, i.e., $Dr \leq 10^5$, which encompasses the domain for practical applications. Nonetheless the maximum value of Dr_c should remain finite. We then compute two portions of the neutral stability curves leading to a central stability window when $k > 5$. The left portion rises dramatically with the Stanton number up to critical draw ratio as high as 10^5 and is in the advection-dominated regime while the right one, located in the heat transfer-dominated regime, strongly decreases with St^* and finally reaches the limit critical draw ratio for $St^* \to \infty$, at which the temperature of the fiber takes the temperature of the ambient. The width of the stability window increases with k as the right portion shift toward the large St^* values while the left portion does not evolve anymore. The production draw ratio being of the order of 10^4 in practical applications, we can see that as soon as $k > 5$ the system admits an operating window compatible with such values, i.e. $Dr_c > 10^4$. In order to obtain a wide enough stable operating window for $Dr \sim 10^4$, and unless specified otherwise, we will take $k = 30$ as reference value in the following sections. Larger values have been found to give similar results (not shown), while bringing numerical difficulties due to the extremely steep temperature gradient it implies near the inlet.

11.4.2 *Influence of Non-constant Heat Transfer Coefficient*

The neutral stability curves evidencing the effect of a non-constant heat transfer coefficient is plotted in Fig. 11.4b. It appears that cooling have a highly stabilizing effect when the heat transfer coefficient follows the Kase and Matsuo model whatever the value of the stiffness parameter. Indeed, just as previously with the largest values of k, the maximum critical draw ratio rises dramatically. Although this value should remain finite, the latter exceeds the limit of the domain covered by the numerical

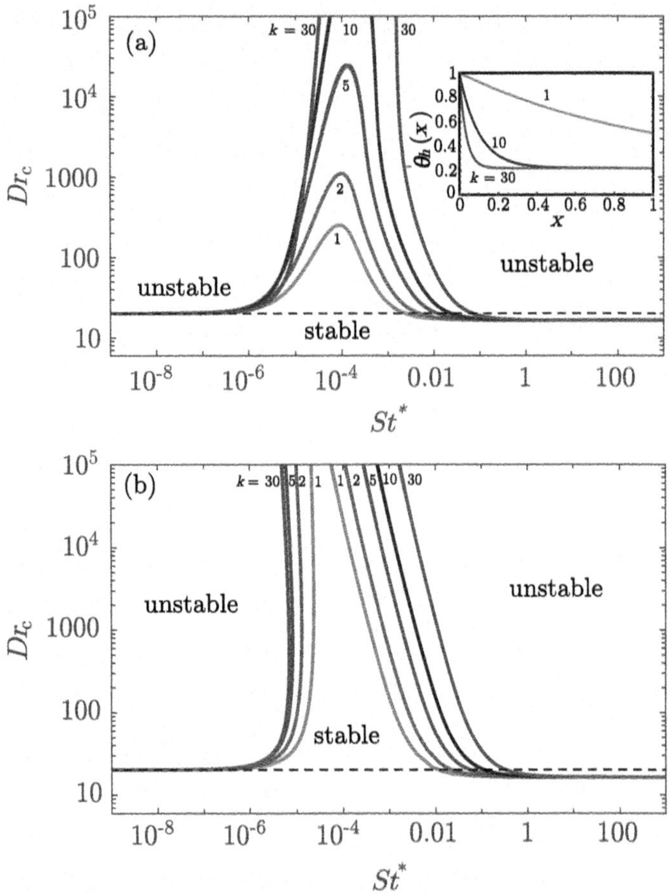

Fig. 11.4 Neutral stability curves for the pure viscous model ($F^* = 0$) with convective heat transfer only ($St_r^* = 0$) and a VFT viscosity law. Each curve is computed for a fiber of aspect ratio $\Lambda = 10^3$, an inlet temperature $\theta_0 = 1573.15$ K and $\theta_{h_\infty} = 0.22$. The black dashed line represents the classical value $Dr_c = 20.218$ encountered in the isothermal case. **a** Influence of the ambient temperature : critical draw ratio as a function of the Stanton number St^* with $k = 1, 2, 5, 10$ and 30 for a constant heat transfer coefficient. The inset represents the ambient temperature $\theta_h(x)$ for $k = 1, 10$ and 30. **b** Influence of the heat transfer coefficient governing law : critical draw ratio as a function of the Stanton number St^* with $k = 1, 2, 5, 10$ and 30 for a heat transfer coefficient following the Kase and Matsuo model (11.10)

computations, i.e., $Dr \leq 10^5$, even for the case $k = 1$. Then, regarding the maximum critical draw ratio, the choice of the stiffness parameter value is less important than the chosen law for the heat transfer coefficient. Nevertheless, the stability window resulting from this new behavior becomes wider as k increases. We also note that for any value of the stiffness parameter the left portion admits a turning point. This important stabilization can be understood by considering the fact that with $a_s u_s$ being constant, the heat transfer coefficient corresponding to the stationary solutions follows the relation $h(a_s, u_s) \propto a_s^{-2/3}$. Then as the cross-section decreases along the fiber, the heat transfer coefficient increases in turn. With the system being in the advection-dominated regime, an increasing heat transfer towards the outlet has a stabilizing effect on the system. It also demonstrates the crucial importance of accounting for a non-constant heat transfer coefficient in thermal fiber drawing.

11.5 Full Non-isothermal Model ($F^* > 0$)

From now on, inertia, gravity, and surface tension are taken into account and we assume that viscosity and heat transfer coefficient follow a VFT law and the Kase and Matsuo model, respectively.

11.5.1 Preliminary Study for the Full Model: Influence of the Fluidity F^*

As seen in Chap. 5, the capillary effect through radial curvature has a dramatic destabilizing effect regarding to draw resonance instability. In order to characterize the effect of the surface tension term (proportional to $F^* S^*$ in the momentum Eq. (11.8b)) in the non-isothermal framework, we first show the influence of the fluidity alone on the stability. Thus we represent in Fig. 11.5 the neutral stability curves obtained for different values of F^* with a dimensionless inlet velocity fixed to $Q^*(\theta_0 = 1603.15K)$ and an aspect ratio $\Lambda = 1000$.

It appears that a dramatic destabilization of the system occurs as the fluidity increases. Indeed, as seen with the pure viscous case, the maximum critical draw ratio is not attainable when the fluidity is low. However, the surface tension has an important destabilizing effect as soon as $F^* = \mathcal{O}(10^{-4})$. It appears that the system admits a regime of unconditional instability as the critical draw ratio can reach unity while its maximum value (if any) decreases at a very strong rate when the fluidity increases. Ultimately the maximum value of Dr_c becomes even lower than 10 when $F^* \approx 3 \times 10^{-3}$. We also note that the Stanton number necessary to attain a stable region increases with F^*, and that the maximum critical draw ratio decreases with F^*. Finally we can see that surface tension effect is also of the higher importance in the non-isothermal framework and that cooling is not enough to stabilize the system

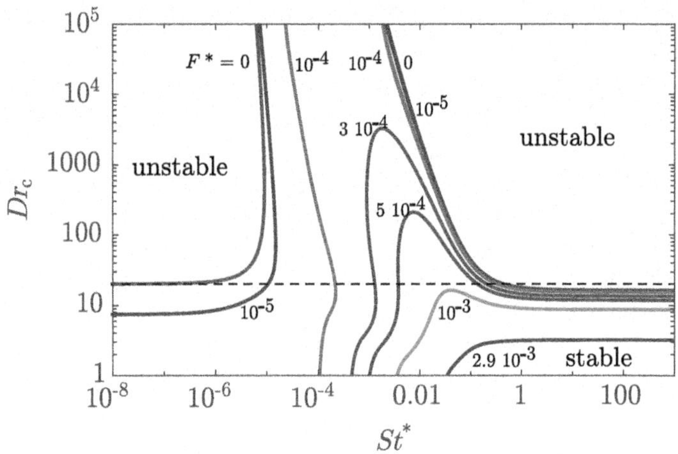

Fig. 11.5 Influence of the fluidity F^*: critical draw ratio as a function of the Stanton number St^* computed for various values of the fluidity F^*, with an aspect ratio $\Lambda = 1000$ and a stiffness parameter $k = 30$. The dimensionless inlet velocity is fixed to its value corresponding to $\theta_0 = 1603.15$ K. The black dashed line represents the classical value $Dr_c = 20.218$ encountered in the pure viscous isothermal case

at any condition. The following sections will be dedicated to the determination of the conditions leading to an important stabilization due to the cooling.

11.5.2 *Influence of the Aspect Ratio* Λ

The neutral stability curves for the full model determined for different values of the fiber aspect ratio are presented in Fig. 11.6 for a fixed inlet temperature.

As long as $\Lambda < 100$, the neutral stability curves admit a maximum out of the domain for practical applications. Similarly to the pure viscous case, we compute two portions of the neutral stability curves leading to a central stability window with critical draw ratio up to 10^5. The left portion admits a turning point at $Dr_c \sim 10$, similarly to the pure viscous case (see Sect. 11.4.2). Hence above this bifurcation point, critical draw ratio increases as the heat transfer decreases. This behavior results in a nonintuitive phenomenon. Indeed, for a fixed value of the Stanton number it is possible to reach a stable region from an unstable one by simply rising the draw ratio. Such a behavior had already been observed both experimentally and numerically for glass fiber drawing (Chouffart, 2018). We also note that the critical draw ratio does not converge to the value of 20.218 corresponding to the pure viscous case as $St^* \to 0$ due to surface tension effects (see Fig. 5.1). And as in the pure viscous case the right portion goes to $Dr_c \approx 16$ as $St^* \to \infty$.

It appears that increasing the control parameter Λ when the inlet temperature is fixed destabilizes the system independently of the value of θ_0. First, the size of the

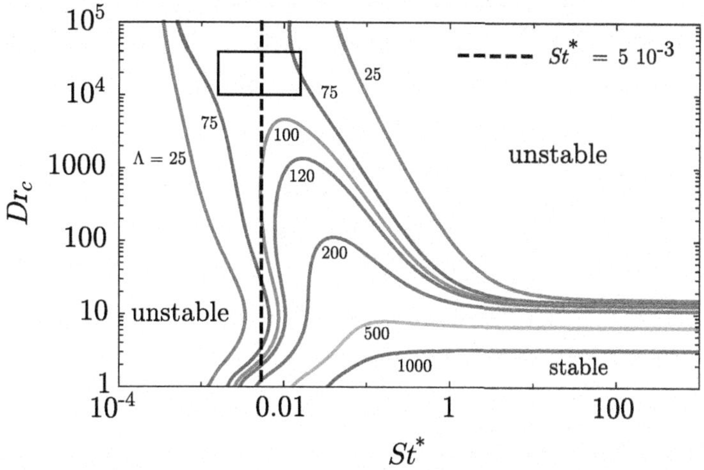

Fig. 11.6 Influence of the fiber aspect ratio: critical draw ratio as a function of the Stanton number St^* computed for fibers of different values of the aspect ratio Λ, with an inlet temperature $\theta_0 =$ 1603.15 K and a stiffness parameter $k = 30$. The range of industrial applications is represented by a black box centered around the realistic Stanton number $St^* = 5 \times 10^{-3}$

stability window decreases as Λ increases, and for $\Lambda \gtrsim 100$, the maximum critical draw ratio decreases dramatically with Λ. Eventually the latter becomes of order unity. The black box drawn in Fig. 11.6 represents a typical range of values of the couple (Dr, St^*) encountered in industrial applications. This box is in a region of strong variability of the critical draw ratio with Λ. We can see that a small variation of the heat transfer coefficient is enough to change the stability of the system for many values of the fiber aspect ratio as exemplified in Fig. 11.6 with the black dashed line $St^* = 5 \times 10^{-3}$. Therefore, the appropriate determination of the fiber aspect ratio to impose in the model is of major interest. Indeed, phase change through glass transition occurs at some region of the fiber and its locus obviously determines the correct value of Λ which remains, in such a framework, an unknown of the problem.

11.5.3 Influence of the Inlet Temperature θ_0 and Stiffness Parameter k

The influence of the inlet temperature θ_0 on the neutral stability curve for the full model encompassing the effects of inertia, gravity and surface tension with a purely convective heat transfer is depicted in Fig. 11.7a.

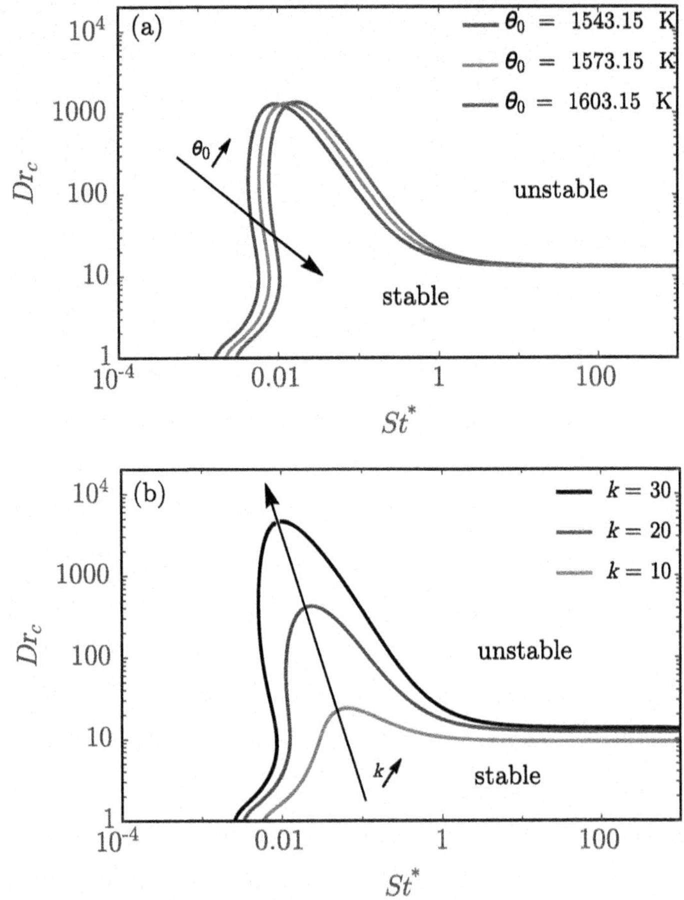

Fig. 11.7 **a** Influence of the inlet temperature on the stability of the system for the full model with convective heat transfer only ($St_r^* = 0$). Critical draw ratio as a function of the Stanton number St^* computed with $F^* = 1.2 \times 10^{-3}$, 1.9×10^{-3} and 2.9×10^{-3} respectively corresponding to inlet temperatures $\theta_0 = 1543.15$, 1573.15 and 1603.15 K. Each curve is obtained for a fiber of aspect ratio $\Lambda = 120$ and a stiffness parameter $k = 30$. **b** Critical draw ratio as a function of the Stanton number St^* computed for stiffness parameter $k = 10$, 20 and 30 with an inlet temperature $\theta_0 = 1603.15$ K and an aspect ratio $\Lambda = 100$

We note that for a given value of the aspect ratio, increasing θ_0 has always a destabilizing effect since the stability curve is shifted toward higher values of the heat transfer while conserving its shape. This behavior is independent of the value of Λ (data not shown). As the inlet temperature controls the fluidity through the viscosity, F^* increases with θ_0. Both inertia and gravity scale with the fluidity and both have a stabilizing effect. However the surface tension parameter is particularly high in our problem. Therefore the surface tension term, which is proportional to

F^*S^* in the momentum Eq. (11.8b) becomes more important and so the system becomes less stable in the low Stanton number regime as θ_0 increases.

The influence of the ambient temperature through the stiffness parameter k on the stability of the system is represented in Fig. 11.7b for a fixed inlet temperature and aspect ratio. Just as in the pure viscous case, we note that the maximum critical draw ratio increases with k and admits a variation of two orders of magnitude between $k = 10$ and $k = 30$. Therefore, a precise depiction of the ambient temperature surrounding the fiber will be of fundamental interest for industrial applications in order to correctly estimate the critical draw ratio. As the stability of the system is very sensitive to ambient temperature and aspect ratio, the temperature of the nozzle θ_0 can be adapted accordingly in order to reach a stable region.

11.5.4 Influence of the Radiative Heat Transfer

As shown by (Huynh and Tanner, 1983), heat exchanges by radiation have an important influence on the first few millimeters below the nozzle and have to be taken into account in the boundary conditions. By fixing ε_0 to a realistic value corresponding to industrial applications (see Table 11.1), the radiative Stanton number is determined only by θ_0.

The influence of the radiative heat transfer is depicted in Fig. 11.8. It compares the neutral stability curves for the full model when radiation is taken into account with those obtained in the purely convective limit. We can see here a stabilizing effect of radiation, especially as $St^* \to 0$. More precisely, stability curves do not reach anymore regions of unconditional instability (i.e. $Dr_c < 1$) with radiation, although the critical draw ratio remains low when the Stanton number becomes very small. Indeed for any value of Λ, $Dr_c \approx 2$ in the low Stanton number limit. We also note that the locus of the turning point occurs at a slightly lower value of the Stanton number.

It is possible to refine the radiative model by introducing a non-constant effective emissivity (Chouffart, 2018), in dimensional form,

$$\varepsilon(a, \theta) = \frac{(a_\varepsilon(\theta_c + \Delta\theta\,\theta) + b_\varepsilon)(1 - e^{c_\varepsilon\sqrt{a_0 a}})}{\varepsilon_0}, \tag{11.19}$$

where $a_\varepsilon = -5.915 \times 10^{-4}\,\mathrm{K}^{-1}$, $b_\varepsilon = 1.227$ and $c_\varepsilon = -2.062 \times 10^4\,\mathrm{m}^{-1}$ are constants depending on the considered molten materials. Contrarily to the heat transfer coefficient, it appears that in the regime we consider, stability curves are not affected by a varying effective emissivity (data not shown).

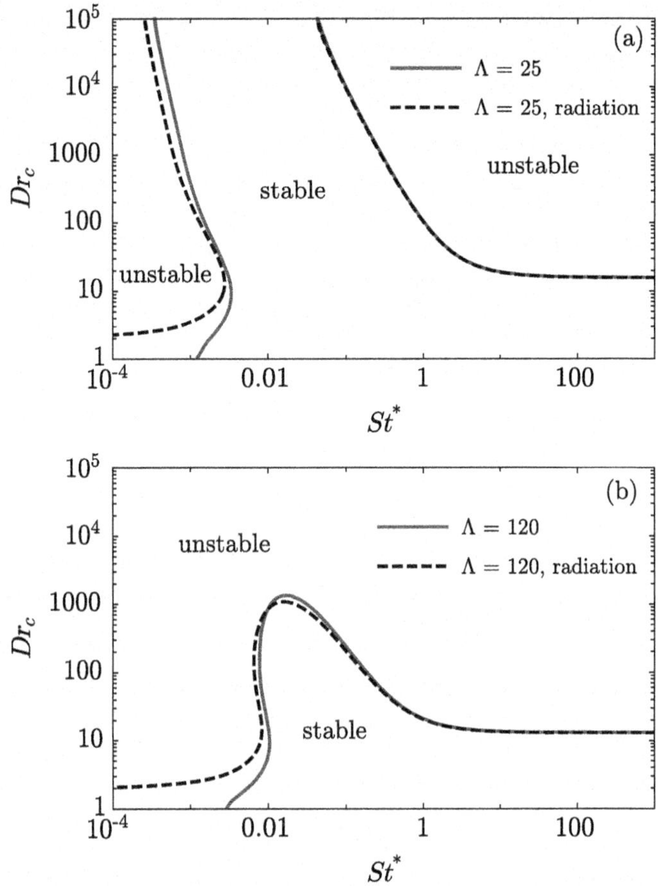

Fig. 11.8 Influence of radiative heat transfer on the stability of the system : Neutral stability curves for the full model represented in the case of a pure convective heat transfer (continuous line) and convective and radiative heat transfer (dashed line) for $\Lambda = 25$ **a** and 120 **b**. Each curve is computed for an inlet temperature $\theta_0 = 1603.15$ K and $k = 30$. The radiative Stanton number is fixed to $St_r^* = 5.8 \times 10^{-5}$. The viscosity and the heat transfer coefficient respectively follow a VFT law and the Kase and Matsuo model

11.6 Few Perspectives

The current model as it is proposed in this chapter could still be improved in several ways. In many industrial applications, the fluid entering in the system through the nozzle simply flows due to gravity. Therefore an hydrostatic pressure is applied at the inlet and a coupling between dynamic pressure $p \sim \eta_0 \partial_x u$, induced by the extensional flow, and velocity fields has to be considered in order to impose a correct boundary condition. Indeed, by considering that the Poiseuille flow in the nozzle is driven by the pressure difference determined by a static and a dynamic contribution, the calculation of the flow rate leads to the following Robin boundary condition :

$$u(0,t) = \frac{qF^*}{\sqrt{\Lambda}} + \frac{R_0}{8\ell\Lambda}\partial_x u(0,t)\ . \tag{11.20}$$

As draw resonance instability is a problem which strongly depends on boundary conditions (Schultz and Davis, 1984), it is expected that such a correction will have an important influence on the stability.

Moreover, the temperature reached in our system due to cooling is always sufficiently high to avoid phase change, namely the glass transition point. However it has been shown for both Newtonian and non-Newtonian fluids that solidification has an influence on the stability (Pearson et al., 1976; Matsumoto and Bogue, 1978; Hagen, 2002). Nevertheless to the best of our knowledge all the physical effects that are relevant in the problem are not taken into account simultaneously in the existing literature. We also note that if the position of the freezing (or glass transition) point is above the winder, the fiber length we have to consider in the system must be modified accordingly. Consequently a full model encompassing phase change has to be developed. A variable fiber length, as proposed here, is a first attempt to deal with this issue, and certainly demonstrates the sensitivity of the stability with respect to this parameter.

By considering the model including inertia, gravity, and surface tension for convective heat transfer only we evidenced that increasing Λ strongly destabilizes the system. We have also demonstrated that the stability of the system is very sensitive to the ambient temperature while a variation of the inlet temperature induces a slight shift of the stability region. These two last parameters can be used to change effectively the critical draw ratio as soon as the heat transfer coefficient is imposed. Therefore the determination of the correct fiber aspect ratio through a model taking into account phase change appears to be a problem of fundamental interest for future research.

Finally, we have shown the limitation of our model due to surface tension effects when the drawing is not sufficient. By adding the effect of longitudinal surface tension, the transition from draw resonance to Rayleigh-Plateau instability could also be the subject of future investigations.

Chapter 12
Conclusions and Outlook

In this book, the influence of process and material parameters on the draw resonance instability in film casting and fiber drawing is investigated both for isothermal and non-isothermal conditions. Draw resonance generally occurs if the draw ratio, i.e., the ratio of outlet to inlet velocity, exceeds a critical value, and manifests itself in steady oscillations of the flow velocity and the geometric properties of the fibers and films. The focus of this book lies hereby on the determination of the critical draw ratio, as this is the parameter of most practical importance.

In order to eliminate the influence of complex material behavior as much as possible, Newtonian and isothermal case are analyzed first. One-dimensional film and fiber models are derived in Chap. 2 in a consistent manner following an expansion in an order parameter. Linear stability analysis is used for determining the critical draw ratio as presented in Chap. 3. Different methods for solving the eigenvalue problem are presented and discussed, and a numerical continuation is chosen as primary procedure, as it enables an efficient calculation of the stability curves in the parameter space. It may fail, however, if the most unstable perturbation mode switches. For this reason, a method based on the full eigenvalue spectrum obtained by Chebyshev collocation is implemented as an alternative. Next, the effects of gravity and inertia, surface tension in fiber drawing and neck-in in film casting are studied separately in Chaps. 4, 5 and 6, respectively.

Gravity, inertia and neck-in are shown to have a purely stabilizing influence, while surface tension purely destabilizes the process. Larger parameter ranges as compared to previous studies are investigated, revealing a region of unconditional stability caused by inertia, and a region of unconditional instability caused by surface tension. Novel, simple empirical stability criteria are given for the influence of gravity, inertia and surface tension and enable an accurate estimation of the critical draw ratio. Based on the results of Kim et al. (1996), the axial force perturbation is identified as the basic quantity for understanding the physical processes underlying the stability

© Springer Nature Switzerland AG 2025

M. Bechert and B. Scheid, *Draw Resonance*, Fluid Mechanics and Its Applications 141,
https://doi.org/10.1007/978-3-031-85568-9_12

effects. Steady state studies for inertia and neck-in effects help additionally to carve out the essential mechanisms.

Chapters 7 and 8 focuses on the influence of viscoelasticity on draw resonance. A fundamental comparative study of common viscoelastic constitutive equations in the scope of draw resonance is still missing in literature. To close this gap, an infinite-width film casting model, neglecting inertia and gravity, is used in order to concentrate on the viscoelastic properties and careful model derivations are given employing the upper-convected Maxwell (UCM) model, the Giesekus model and the simplified Phan-Thien–Tanner (PTT) model. The critical draw ratio in dependence of the Deborah number is calculated for all three models. Apart from the UCM model, which is not recommended for draw resonance studies, all models yield a mostly exponentially decaying critical draw ratio with increasing Deborah number, together with a small stability maximum if the nonlinear model parameter is in an appropriate range. A shifted power-law can be used as an empirical stability criterion for Deborah numbers above 0.1. Evaluation of the stability results yields the PTT model as the most appropriate candidate for further modeling, even though perturbation mode switching is observed for values of the nonlinear parameter ε above or equal to 0.5.

The effective elongational viscosity is identified as a crucial quantity and examined for the first time in the scope of draw resonance. Mastercurves independent of the draw ratio are revealed and exhibit strong correlation to the neutral stability curves. Motivated by this finding, generalized Newtonian fluid (GNF) models are tested for their ability to reproduce the results of the viscoelastic models. While the steady state predictions are in good agreement, the stability results coincide only for small values of the Deborah number, which indicates that elastic effects, which are not covered by GNF models, influence also the draw resonance behavior. A deeper understanding of these elastic effects is still missing, though. An analysis of a GNF model based on the Carreau–Yasuda function reveals the stability mechanism underlying the effects of strain hardening and thinning, which is governed by an interplay of a static and a dynamic mechanism. Kim's stability criterion is checked for both Newtonian and viscoelastic models and can always be verified. Jung's approximative criterion (Jung et al., 2000), however, is strongly limited and hardly extendible to cover additional effects in a feasible way.

More realistic models accounting for several effects simultaneously are analyzed in Chap. 9. An alternative scaling yields dimensionless control parameters with strong connection to practical applications. Novel stability maps with isolines of the critical draw ratio visualize the stability behavior in the practical relevant parameter space for various Newtonian and viscoelastic models. Moreover, they enable the identification of several dynamical regimes, within which one or several effects can be neglected, together with regions of unconditional stability or instability. By calculating the relevant control parameters of a particular process, appropriate model assumptions can easily be determined using these maps. Trajectories of special importance for practical applications, e.g., the threshold to unconditional stability or local stability maxima, are tracked additionally in the parameter space. In addition to the stability maps, which pose an efficient tool for quickly estimating the critical draw ratio, empirical stability criteria are given for various dynamical regimes in the cases of infinite-width film casting and fiber drawing with negligible surface tension.

The last two chapters concern the investigation of draw resonance under non-isothermal conditions. Because of the added-complexity of including the temperature field, only Newtonian fluids have been considered, which is full compatible to process invoving molten glass such as the float glass and the glass fiber drawing.

In Chap. 10 we have investigated the stability of a stretching viscous sheet for the entire range of Stanton numbers (St) and have identified two regimes: one dominated by advection of energy as is commonly found in the literature for fiber drawing and film casting whose process parameters typically correspond to $St \lesssim 1$, and one dominated by heat transfer to the environment as encountered for instance in the glass sheet industry whose process parameters typically correspond to $St \gg 1$. While heat transfer has a stabilizing effect in the first regime, it has a destabilizing effect in the second one where the liquid sheet approaches the imposed temperature of the surrounding as St increases. To some extent, this regime can be associated with a cooling dominated by radiative heat transfer rather than by convective heat transfer, as is the case for manufacturing glass sheets. This conclusion is supported by the work by Willien et al. (1988) who found in fiber drawing that radiative heat transfer can be destabilizing, which is opposite to the response for the case where convective heat transfer is dominant.

We also propose a mechanism that qualitatively explains the stabilizing effect due to cooling, which is most efficient in the region where the energy advected by the flow and energy transferred to the ambient are comparable, i.e. for $St = \mathcal{O}(1)$. In this region, the average viscosity is shown to act quasi-instantaneously against a perturbation of the tension, as a consequence of an in-phase variation of the axial-averaged flow rate and axial-averaged temperature. In the limit of $St \to \infty$, the temperature of the sheet is shown to be prescribed by the temperature of the ambient environment, which leads to a useful simplification where the viscosity is a prescribed function of the coordinates. Nevertheless, we have to remember that at finite St, the temperature field always stabilizes the system compared to the infinite St limit. This limit represents then the 'worst case', where the critical draw ratio can be below its value for isothermal conditions when increasing the temperature sensitivity of the viscosity, at least in the case of symmetrical thermal boundary conditions. In the case of non-symmetrical thermal boundary conditions, the present model would also account for viscosity variation in the transverse direction. This subject is one for future study.

In Chap. 11, we have investigated the stability of Newtonian fiber drawing processes with an emphasis on molten glass in the non-isothermal case. Our one-dimensional model generalizes the one proposed in Chap. 2, which encompasses simultaneously the effects of inertia, gravity and surface tension, by adding the cross-section averaged energy equation for convective and radiative heat transfer. The system is closed through the VFT law which characterizes the temperature-dependant viscosity. With the length L of the system being not always very well defined, and should depend on the location of the glass transition point, we have chosen to consider it as a new control parameter of the problem via the aspect ratio and have used the rescaling introduced in Chap. 9. Finally the fiber drawing model for molten glass we have proposed is governed by five control parameters, namely the draw ratio Dr, the inlet temperature θ_0, the Stanton number St^*, the fiber aspect ratio Λ and the stiffness k of the ambient temperature decay. By the use of the linear stability analysis we have first shown in the pure viscous regime that the strong stabilization induced by cooling is not related to the law of viscosity we consider but is rather due to the modelling of the heat transfer coefficient of the fiber through the Kase and Matsuo model. Within this framework we obtained a model explaining how it is possible to reach critical draw ratio as high as 10^4 as encountered in industrial applications.

Correction to: Draw Resonance

Correction to:
M. Bechert and B. Scheid, *Draw Resonance*, Fluid Mechanics and Its Applications 141,
https://doi.org/10.1007/978-3-031-85568-9

The original version of the series volume number has been updated from 990 to 141. The book has been updated with the changes.

The updated version of this book can be found at
https://doi.org/10.1007/978-3-031-85568-9

© Springer Nature Switzerland AG 2025

M. Bechert and B. Scheid, *Draw Resonance*, Fluid Mechanics and Its Applications 141,
https://doi.org/10.1007/978-3-031-85568-9_13

Appendix A
Eigenvalue Spectrum Determination

In most cases throughout this book, a numerical continuation method (see Sect. 3.2.2) and Appendix B) is used to track the most unstable mode in the control parameter space. Sometimes, however, it is worth studying the whole eigenvalue spectrum, as the continuation method can be erroneous[1] (see for instance Sects. 8.1.2 and 8.1.3). This can be achieved by a discretization of the eigenvalue problem, which leads to the generalized eigenvalue problem (3.27). As both matrices $\boldsymbol{A}$ and $\boldsymbol{B}$ depend on the steady state solutions in general, these need to be found prior to solving the eigenvalue problem. Here, a shooting method is utilized for solving the boundary value problem formed with the steady state equations together with the boundary conditions. This method transforms the boundary value problem into an initial value problem by guessing an additional initial boundary condition. Here, the normal stress difference at the inlet is specified. This condition is iteratively adjusted and with each step, the system is solved by means of numerical integration, until the missing boundary condition is fulfilled, i.e., $u_s(1) = Dr$. The ODEINT method of the Python package SCIPY can be used to perform the numerical integration. While the shooting method has a couple of drawbacks, e.g., it is not sure whether the desired boundary condition is approached continuously by changing the guessed initial condition, it benefits from the advantage of well-established numerical methods for solving initial value problems and it appeared to yield reliable and fast results in the context of this book. Following Hagen and Langwallner (2006), a Chebyshev collocation method can be used for solving the eigenvalue problem. A detailed overview of the method and general properties of the Chebyshev polynomials can be found in the textbook of Peyret (2002). The general idea is to discretize the perturbations, e.g., $U(x)$, not by using an equally spaced grid, but by expansion in terms of the Chebyshev polynomials of first kind, $T_i(x)$, e.g.,

[1] Indeed, the continuation method is unable to detect a switch of the modes, unless all modes are continued individually. It can thus results in missing the most unstable mode.

© Springer Nature Switzerland AG 2025

M. Bechert and B. Scheid, *Draw Resonance*, Fluid Mechanics and Its Applications 141,
https://doi.org/10.1007/978-3-031-85568-9

$$U(x) = \sum_{i=0}^{J} \hat{U}_i T_i(2x-1) \,, \tag{A.1}$$

with $\hat{U}_i$ being the $J+1$ Chebyshev coefficients. The equations are then discretized by evaluating them at the collocation points $\hat{x}_i$, at which the Chebyshev polynomials reach their extremal values ± 1, and which are defined as

$$\hat{x}_i = \frac{1}{2}\left(\cos\left(\frac{i\pi}{J}\right) + 1\right), \; 0 \le i \le J \,. \tag{A.2}$$

Note that the expansion in the Chebyshev polynomials is shifted so that the collocation points are bounded by 0 and 1, the domain of the system.

Due to the properties of the Chebyshev polynomials, the derivative of the expanded and discretized quantities can be calculated by transforming the Chebyshev coefficients with the matrix $\boldsymbol{D}$, defined as

$$(\boldsymbol{D})_{ij} = \frac{4}{c_i} d_{ij} \,, \tag{A.3}$$

with

$$c_i = \begin{cases} 2 & \text{if } i = 0 \\ 1 & \text{if } i \ge 1 \end{cases} \,, \tag{A.4a}$$

$$d_{ij} = \begin{cases} 0 & \text{if } j \le i \\ 0 & \text{if } i = J \\ 0 & \text{if } (i+j) \text{ even} \\ j & \text{else} \end{cases} \,. \tag{A.4b}$$

The vector of Chebyshev coefficients can be transformed to the vector of the discretized quantity by applying the matrix $\boldsymbol{M}$, defined as

$$(\boldsymbol{M})_{ij} = \cos\left(\frac{i\pi j}{J}\right) \,. \tag{A.5}$$

The matrices $\boldsymbol{A}$ and $\boldsymbol{B}$ can be constructed with the help of $\boldsymbol{M}$ and $\boldsymbol{D}$, substituting single lines to define the boundary conditions. The eigenvalue problem itself is then solved with the EIGVALS method of the Python package SCIPY and numerically caused infinite values, zero imaginary part eigenvalues and spurious eigenmodes (van der Walt et al. 2012) are eliminated afterwards. The great advantage of discretizing the system at the Chebyshev collocation points is that the eigenvalue determination converges already for $J \gtrsim 100$, while a common equally spaced discretization yields very imprecise values with the same number of mesh points.

Appendix B
Continuation Method Using AUTO-07P: Tutorial

We show in this tutorial the procedure to compute a stability curve for draw resonance by continuation. It relies on the assumption that the solution we are looking for, namely the neutral mode, is not isolated in the parameter space and lies on a continuous branch of solutions when varying one single parameter. One can then construct the whole branch of solutions in small steps, starting from a known solution. An accurate continuation algorithm has been developed by Doedel et al. (2012) and implemented in the software package AUTO-07P which can be downloaded from http://indy.cs.concordia.ca/auto/. The software is designed as a collection of subroutines that enables the solution of bifurcation problems for ordinary differential equations. It can be installed on most operating systems, like Windows, even though it is more straightforward on Unix-based ones, including Linux and Mac. For more details, we refer the reader to the documentation of the software. Here we only provide the basic steps to using the software as it has been used to generate all stability curves in this book.

In this tutorial, we show as an example the steps to build the stability curve presented in Fig. 4.5 to account for the effect of inertia on draw resonance. We provide as supplemental materials all the files necessary to run this example using AUTO-07P, which are the equation file `dra.f`, the constant files `c.dra` and `c.dra.R`, and the data file `dra.dat` containing the initial solution. This solution has been obtained by solving (3.15) and then (3.17) — using ©Mathematica for instance — for a draw ratio in the stable regime $Dr < Dr_c$. We have arbitrarily chosen the solution for $Dr = 16$, with $\omega_R = -0,494223$ and $\omega_I = 13.27843$ found with the characteristic equation (3.18).

The procedure is as follows:

Step 0. Open a terminal and go in the directory that contains the files.

Step 1. Convert the data file `dra.dat` into a starting solution `s.dat` for AUTO-07P with the command:

© Springer Nature Switzerland AG 2025

M. Bechert and B. Scheid, *Draw Resonance*, Fluid Mechanics and Its Applications 141,
https://doi.org/10.1007/978-3-031-85568-9

```
@fc dra
```

Step 2. Using the draw ratio as the principal continuation parameter (`PAR(1)`), increase its value up to the neutral mode (`PAR(2)=0`) that arises for `PAR(1)= 20.218`, which is recorded as `LAB 2`, and save the solution into the solution files `b.dra`, `s.dra` and `d.dra`:

```
@r dra dat
@sv dra
```

Step 3. Using now the Reynolds number as the principal continuation parameter (`PAR(5)`), increase the Reynolds number from the previous solution obtained with `PAR(5)=0`, and append the solution to the solution files:

```
@r dra R
@ap dra
```

Step 4. Plot the results and choose `PAR(5)` for the `X` axis and `PAR(1)` for the `Y` axis such as to obtain the same curve as the one in Fig. 4.5.

```
@pp dra
```

The above procedure has been applied for computing all the neutral stability curves presented in this book. It relies on the assumption that the most amplified mode (3.19) used as the starting point remains the most amplified mode in the whole parameter space. As explained in Sect. 3.2.3, there is however no guarantee that this is the case and checks with other methods like the one presented in Appendix A has to be done.

References

Agassant, J., Avenas, P., Carreau, P., Vergnes, B., Vincent, M.: Polymer Processing: Principles and Modeling. Carl Hanser Verlag GmbH & Company KG (2017). ISBN 978-1-56990-606-4

Ahmed, Z.U., Khayat, R.E.: Three dimensional film stability and draw resonance. J. Fluids Eng. **134**(10), 101302–101302 (2012)

Aird, G.R., Yeow, Y.L.: Stability of film casting of power-law liquids. Ind. Eng. Chem. Fund. **22**(1), 7–10 (1983)

Alaie, S.M., Papanastasiou, T.C.: Film casting of viscoelastic liquid. Polymer Eng. Sci. **31**(2), 67–75 (1991)

Barot, G., Rao, I.J.: Modeling the film casting process using a continuum model for crystallization in polymers. Int. J. Nonlinear Mech. **40**, 939 (2004)

Barq, P., Haudin, J.M., Agassant, J.F., Roth, H., Bourgin, P.: Instability phenomena in film casting process. Int. Polymer Process. **5**(4), 264–271 (1990)

Barq, P., Haudin, J.M., Agassant, J.F., Bourgin, P.: Stationary and dynamic analysis of film casting process. Int. Polymer Process. **9**(4), 350–358 (1994)

Beaulne, M., Mitsoulis, E.: Numerical simulation of the film casting process. Int. Polymer Process. **14**(3), 261–275 (1999)

Bechert, M.: Non-newtonian effects on draw resonance in film casting. J. Non-Newtonian Fluid Mech. **279**, 104262 (2020)

Bechert, M., Scheid, B.: Combined influence of inertia, gravity and surface tension on the linear stability of Newtonian fiber spinning. Phys. Rev. Fluids **2**, 113905 (2017)

Bechert, M., Schubert, D.W., Scheid, B.: Practical mapping of the draw resonance instability in film casting of Newtonian fluids. Eur. J. Mech.–B/Fluids **52**, 68–75 (2015)

Bechert, M., Schubert, D.W., Scheid, B.: On the stabilizing effects of neck-in, gravity, and inertia in Newtonian film casting. Phys. Fluids **28**, 024109 (2016)

Bergonzoni, A., DiCresce, A.J.: The phenomenon of draw resonance in polymeric melts. Part I-qualitative view. Polymer Eng. Sci. **6**(1), 45–49 (1966)

Bergonzoni, A., DiCresce, A.J.: The phenomenon of draw resonance in polmeric melts. Part II–correlation to molecular parameters. Polymer Eng. Sci. **6**(1), 50–59 (1966b)

Beyreuther, R., Brünig, H.: Dynamics of fibre formation and processing: modelling and application in fibre and textile industry. Springer (2007)

Blyler, L.L., Gieniewski, C.: Melt spinning and draw resonance studies on a poly (α-methyl styrene/silicone) block copolymer. Polymer Eng. Sci. **20**(2), 140–148 (1980)

© Springer Nature Switzerland AG 2025

M. Bechert and B. Scheid, *Draw Resonance*, Fluid Mechanics and Its Applications 141,
https://doi.org/10.1007/978-3-031-85568-9

Bouamra, R., Agassant, J.-F.O., Peiti, C., Haudin, J.M.: Stability of cast-film extrusion of various metallocene polyethylenes. In: 22nd Annual Meeting of the Polymer Processing Society (PPS-22) (2006)

Burghelea, T.I., Grieß, H.J., Münstedt, H.: An in situ investigation of the draw resonance phenomenon in film casting of a polyproylene melt. J. Non-Newtonian Fluid Mech. **173–174**, 87–96 (2012)

Cao, F., Khayat, R.E., Puskas, J.E.: Effect of inertia and gravity on the draw resonance in high-speed film casting of Newtonian fluids. Int. J. Solids Struct. **42**(21–22), 5734–5757 (2005)

Chandrasekhar, S.: Hydrodynamic and hydromagnetic stability. Clarendon Press, Oxford (1961)

Chang, J.C., Denn, M.M.: An experimental study of isotermal spinning of a Newtonian and viscoelastic liquid. J. Non-Newtonian Fluid Mech. **5**, 369–385 (1979)

Chang, J.C., Denn, M.M., Geyling, F.T.: Effects of inertia, surface tension, and gravity on the stability of isothermal drawing of Newtonian fluids. Ind. Eng. Chem. Fund. **20**(2), 147–149 (1981)

Chang, J.-C., Denn, M.M., Kase, S.: Dynamic simulation of low-speed melt spinning. Ind. Eng. Chem. Fund. **21**(1), 13–17 (1982)

Charru, F.: Hydrodynamic Instabilities. Cambridge Texts in Applied Mathematics. Cambridge University Press (2011)

Chen, Z., Papanastosiou, A.C.: Fiber spinning with molecular models. Int. Polymer Process. **2**(1), 33–40 (1987)

Chouffart, Q.: Experimental and numerical investigation of the continuous glass fiber drawing process. PhD thesis, Université de Liège (2018)

Chouffart, Q., Simon, P., Terrapon, V.E.: Numerical and experimental study of the glass flow and heat transfer in the continuous glass fiber drawing process. J. Materials Process. Technol. **231**, 75–88 (2016)

Christensen, R.E.: Extrusion coating of polypropylene. SPE J. 751–755 (1962)

Christodoulou, K., Hatzikiriakos, S.G., Vlassopoulos, D.: Stability analysis of film casting for PET resins using a multimode phan-thien-tanner constitutive equation. J. Plastic Film Sheet. **16**(4), 312–332 (2000)

Co, A.: Draw resonance in film casting. In: Hatzikiriakos, S.G., Migler, K.B. (eds.) Polymer Processing Instabilities: Control and Understanding. CRC Press, New York (2004)

Dantzig, J.A., Tucker, C.L.: Modeling in Materials Processing. Cambridge University Press, Cambridge (2001)

Dealy, J.M., Larson, R.G.: Structure and Rheology of Molten Polymers: From Structure to Flow Behavior and Back Again. Carl Hanser Verlag, München (2006)

Demay, Y., Agassant, J.-F.: Experimental study of the draw resonance in fiber spinning. J. Non-Newtonian Fluid Mech. **18**(2), 187–198 (1985)

Demay, Y., Agassant, J.-F.: An overview of molten polymer drawing instabilities. Int. Polymer Process. **29**(1), 128–139 (2014)

Denn, M.: Polymer Melt Processing: Foundations in Fluid Mechanics and Heat Transfer. Cambridge Series in Chemical Engineering. Cambridge University Press (2008). ISBN 978-1-316-58314-2

Denn, M.M.: Continuous drawing of liquids to form fibers. Ann. Rev. Fluid Mech. **12**(1), 365–387 (1980)

Denn, M.M.: On an analysis of draw resonance by Hyun. AIChE J. **26**(2), 292–294 (1980)

Denn, M.M., Petrie, C.J., Avenas, P.: Mechanics of steady spinning of a viscoelastic liquid. AIChE J. **21**(4), 791–799 (1975)

Dewynne, J., Ockendon, J.R., Wilmott, P.: On a mathematical model for fiber tapering. SIAM J. Appl. Math. **49**(4), 983–990 (1989)

Dewynne, J.N., Ockendon, J.R., Wilmott, P.: A systematic derivation of the leading-order equations for extensional flows in slender geometries. J. Fluid Mech. **244**, 323–338 (1992)

Dhadwal, R., Banik, S., Doshi, P., Pol, H.: Effect of viscoelastic relaxation modes on stability of extrusion film casting process modeled using multi-mode Phan-Thien-Tanner constitutive equation. Appl. Math Model. **47**, 487–500 (2017) d'Halewyu, S., Agassant, J.F., Demay, Y.: Numerical simulation of the cast film process. Polymer Eng. Sci. **30**(6), 335–340 (1990)

Dobroth, T., Erwin, L.: Causes of edge beads in cast films. Polymer Eng. Sci. **26**(7), 462–467 (1986)

Doedel, E., Oldeman, B.E., Champneys, A.R., Dercole, F., Fairgrieve, T., Kuznetsov, Y., Paffenroth, R., Sandstede, B., Wang, X., Zhang, C.: AUTO-07P: Continuation and Bifurcation Software for Ordinary Differential Equations (2012). AUTO software is freely distributed on http://indy.cs.concordia.ca/auto/

Donnelly, G.J., Weinberger, C.B.: Stability of isothermal fiber spinning of a Newtonian fluid. Ind. Eng. Chem. Fund. **14**(4), 334–337 (1975)

Drazin, P.G., Reid, W.H.: Hydrodynamic Stability, 2nd edn. Cambridge University Press, Cambridge Mathematical Library (2004)

Fisher, R.J., Denn, M.M.: Finite-amplitude stability and draw resonance in isothermal melt spinning. Chem. Eng. Sci. **30**(9), 1129–1134 (1975)

Fisher, R.J., Denn, M.M.: A theory of isothermal melt spinning and draw resonance. AIChE J. **22**(2), 236–246 (1976)

Fisher, R.J., Denn, M.M.: Mechanics of nonisothermal polymer melt spinning. AIChE J. **23**(1), 23–28 (1977)

Forest, M.G., Zhou, H., Wang, Q.: Thermotropic liquid crystalline polymer fibers. SIAM J. Appl. Math. **60**(4), 1177 (2000)

Fourné, F.: Synthetic Fibers: Machines and Equipment, Manufacture. Machine Design, and Operation. Hanser, Properties, Handbook for Plant Engineering (1999)

Fulchiron, R., Michel, A., Verney, V., Roustant, J.C.: Correlations between relaxation time spectrum and melt spinning behavior of polypropylene. II: Melt spinning simulation from relaxation time spectrum. Polymer Eng. Sci. **35**(6), 518–527 (1995a)

Fulchiron, R., Michel, A., Verney, V., Roustant, J.C.: Correlations between relaxation time spectrum and melt spinning behavior of polypropylene. 1: Calculation of the relaxation spectrum as a log-normal distribution and influence of the molecular parameters. Polymer Eng. Sci. **35**(6), 513–517 (1995b)

Fulchiron, R., Revenu, P., Kim, B.S., Carrot, C., Guillet, J.: Extrudate swell and isothermal melt spinning analysis of linear low density polyethylene using the Wagner constitutive equation. J. Non-Newtonian Fluid Mech. **69**(2–3), 113–136 (1997)

Gallaire, F., Brun, P.-T.: Fluid dynamic instabilities: theory and application to pattern forming in complex media. Phil. Trans. R. Soc. A. **375**, 20160155 (2017)

Gelder, D.: The stability of fiber drawing processes. Ind. Eng. Chem. Fund. **10**(3), 534–535 (1971)

German, R., Khayat, R.E.: Interplay between inertia and elasticity in film casting. J. Fluids Eng. **130**(8), 081501–081501 (2008). ISSN 0098-2202

German, R., Khayat, R.E., Cui, J.K.: Influence of inertia and gravity on the stability of filament jet flow. Phys. Fluids **18**(6), 064108 (2006)

Geyling, F.T., Homsy, G.M.: Extensional instabilities of glass fibre drawing process. Glass Tech. **21**, 95–102 (1980)

Ghijsels, A., Ente, J.J.S.M.: Draw Resonance Studies on Polypropylene Melts. In: Astarita, P.G., Marrucci, P.G., Nicolais, P.L. (eds.) Rheology, pp. 15–24. Springer, US (1980)

Grieß, H.J.: Fließverhalten von Polypropylenschmelzen innerhalb und außerhalb einer Kleiderbügeldüse. PhD thesis, Friedrich-Alexander-Universität Erlangen-Nürnberg (2014)

Gupta, G.K., Schultz, W.W., Arruda, E.M., Lu, X.: Nonisothermal model of glass fiber drawing stability. Rheologica Acta **35**(6), 584–596 (1996)

Gupta, K., Chokshi, P.: Weakly nonlinear stability analysis of polymer fibre spinning. J. Fluid Mech. **776**, 268–289 (2015)

Gupta, K., Chokshi, P., Stepanyan, R.: Die-swell effect in draw resonance of polymeric spin-line. J. Non-Newtonian Fluid Mech. **230**, 1–11 (2016)

Hagen, T.: Eigenvalue Asymptotics in Isothermal Forced Elongation. J. Math. Anal. Appl. **244**(2), 393–407 (2000)

Hagen, T.: On the effects of spinline cooling and surface tension in fiber spinning. ZAMM–J. Appl. Math. Mech. Zeitschrift für Angewandte Mathematik und Mechanik, **82**(8), 545–558 (2002)

Hagen, T., Langwallner, B.: A numerical study on the suppression of draw resonance by inertia. ZAMM **86**(1), 63–70 (2006)

Hagen, T., Renardy, M.: Eigenvalue asymptotics in non-isothermal elongational flow. J. Math. Anal. Appl. **252**(1), 431–443 (2000)

Hatzikiriakos, S.G., Migler, K.B. (eds.): Polymer Processing Instabilities: Understanding and Control. CRC Press, New York (2004)

Huynh, B.P., Tanner, R.I.: Study of the non-isothermal glass fibre drawing process. Rheologica Acta **22**(5), 482–499 (1983)

Hyun, J.C.: Theory of Draw Resonance: Part I. Newtonian Fluids. AIChE J. **24**(3), 418–422 (1978)

Hyun, J.C.: Rebuttal of Denn's note. AIChE J. **26**(2), 294–295 (1980)

Hyun, J.C.: Draw resonance in polymer processing: a short chronology and a new approach. Korea-Australia Rheology J. **11**, 279–285 (1999)

Ito, H., Doi, M., Isaki, T., Takeo, M.: A model of neck-in phenomenon in film casting process. Nihon Reoroji Gakkaishi **31**(3), 157–163 (2003)

Iyengar, V.R., Co, A.: Film casting of a modified Giesekus fluid: a steady-state analysis. J. Non-Newtonian Fluid Mech. **48**, 1–20 (1993)

Iyengar, V.R., Co, A.: Film casting of a modified Giesekus fluid: Stability analysis. Chem. Eng. Sci. **51**(9), 1417–1430 (1996)

Jung, H.W.: Fiber spinning and film blowing instabilities. In: Hatzikiriakos, S.G., Migler, K.B. (eds.) Polymer Processing Instabilities: Control and Understanding. CRC Press, New York (2004)

Jung, H.W., Hyun, J.C.: Instabilities in extensional deformation polymer processing. Rheology Reviews **2006**, 131 (2006)

Jung, H.W., Choi, S.M., Hyun, J.C.: Approximate method for determining the stability of the film-casting process. AIChE J. **45**(5), 1157–1160 (1999)

Jung, H.W., Song, H.-S., Hyun, J.C.: Draw resonance and kinematic waves in viscoelastic isothermal spinning. AIChE J. **46**(10) (2000)

Kallel, A., Hachem, E., Rapetti, F., Demay, Y., Agassant, J.F.: Stability analysis of a polymer film casting problem. Int. J. Numerical Methods Fluids **78**(7), 436–454 (2015)

Kase, S.: Studies on melt spinning. IV. On the stability of melt spinning. J. Appl. Polymer Sci. **18**(11), 3279–3304 (1974)

Kase, S., Matsuo, T.: Studies on melt spinning. I. Fundamental equations on the dynamics of melt spinning. J. Polymer Sci. Part A: General Papers **3**(7), 2541–2554 (1965)

Kim, B.M., Hyun, J.C., Oh, J.S., Lee, S.J.: Kinematic waves in the isothermal melt spinning of Newtonian fluids. AIChE J. **42**(11), 3164–3169 (1996)

Kim, J.M., Lee, J.S., Shin, D.M., Jung, H.W., Hyun, J.C.: Transient solutions of the dynamics of film casting process using a 2-D viscoelastic model. J. Non-Newtonian Fluid. Mech. **132**(1–3), 53–60 (2005)

Landau, L.D., Lifshitz, E.M.: Statistical Physics. Pergamon Press, Oxford (1980)

Larson, R.G.: Constitutive Equations for Polymer Melts and Solutions. Butterworth, Boston (1988)

Larson, R.G.: Instabilities in viscoelastic flows. Rheologica Acta **31**(3), 213–263 (1992)

Lee, J.S., Hyun, J.C.: Nonlinear dynamics and stability of film casting process. Korea-Aust. Rheo. J. **13**(4), 179–187 (2001)

Lee, J.S., Jung, H.W., Kim, S.H., Hyun, J.C.: Effect of fluid viscoelasticity on the draw resonance dynamics of melt spinning. J. Non-Newtonian Fluid. Mech. **99**(2), 159–166 (2001)

Lee, J.S., Jung, H.W., Song, H.-S., Lee, K.-Y., Hyun, J.C.: Kinematic waves and draw resonance in film casting process. J. Non-Newtonian Fluid Mech. **101**(1–3), 43–54 (2001)

Lee, J.S., Jung, H.W., Hyun, J.C., Scriven, L.E.: Simple indicator of draw resonance instability in melt spinning processes. AIChE J. **51**(10), 2869–2874 (2005)

Lee, J.S., Shin, D.M., Jung, H.W., Hyun, J.C.: Stability analysis of a three-layer film casting process. Korea-Aust. Rheo. J. **19**(1), 27–33 (2007)

Macosko, C.W.: Rheology: Principles, Measurements and Applications. Wiley-VCH, New York (1994)

Manneville, P.: Instabilities. Chaos and Turbulence. Imperial College Press, London (2010)

Matovich, M.A., Pearson, J.R.A.: Spinning a molten threadline. Steady-state isothermal viscous flows. Ind. Eng. Chem. Fund. **8**(3), 512–520 (1969)

Matsumoto, T., Bogue, D.C.: Draw resonance involving rheological transitions. Polymer Eng. Sci. **18**(7), 564–571 (1978)

Miller, J.C.: Swelling behavior in extrusion. Polymer Eng. Sci. **3**(2), 134–137 (1963)

Mitsoulis, E., Beaulne, M.: Numerical simulation of rheological effects in fiber spinning. Adv. Polym. Technol. **19**(3), 155–172 (2000)

Nam, S., Bogue, D.C.: Dynamics of steady and unsteady melt spinning. Ind. Eng. Chem. Fundam. **23**(1), 1–8 (1984)

Narayanaswamy, O.S.: A one-dimensional model of stretching float glass. J. Am. Ceram. Soc. **60**(1–2), 1–5 (1977)

Papanastasiou, T.C., Macosko, C.W., Scriven, L.E., Chen, Z.: Fiber spinning of viscoelastic liquid. AIChE J. **33**(5), 834–842 (1987)

Papanastasiou, T.C., Dimitriadis, V.D., Scriven, L.E., Macosko, C.W., Sani, R.L.: On the inlet stress condition and admissibility of solution of fiber-spinning. Adv. Polym. Technol. **15**(3), 237–244 (1996)

Pearson, J.R.A.: Mechanics of Polymer Processing. Elsevier Applied Science Publishers (1985)

Pearson, J.R.A., Matovich, M.A.: Spinning a molten threadline. Stability. Ind. Eng. Chem. Fund. **8**(4), 605–609 (1969)

Pearson, J.R.A., Shah, Y.T.: On the stability of isothermal and nonisothermal fiber spinning of power-law fluids. Ind. Eng. Chem. Fundam. **13**(2), 134–138 (1974)

Pearson, J.R.A., Shah, Y.T., Mhaskar, R.D.: On the stability of fiber spinning of freezing fluids. Ind. Eng. Chem. Fundam. **15**(1), 31–37 (1976)

Petrie, C.J., Denn, M.M.: Instabilities in polymer processing. AIChE J. **22**(2), 209–236 (1976)

Peyret, R.: Spectral Methods for Incompressible Viscous Flow. Springer, New York (2002)

Pilkington, L.A.B.: Review lecture. The float glass process. Proc. T. Soc. London, Ser. A, **314**, 1–25 (1969)

Renardy, M.: Draw Resonance Revisited. SIAM J. Appl. Math. **66**(4), 1261–1269 (2006)

Sakaki, K., Katsumoto, R., Kajiwara, T., Funatsu, K.: Three-dimensional flow simulation of a film-casting process. Polymer Eng. Sci. **36**(13), 1821–1831 (1996)

Satoh, N., Tomiyama, H., Kajiwara, T.: Viscoelastic simulation of film casting process for a polymer melt. Polymer Eng. Sci. **41**(9), 1564–1579 (2001)

Scheid, B., Quiligotti, S., Tran, B., Gy, R., Stone, H.A.: On the (de)stabilization of draw resonance due to cooling. J. Fluid Mech. **636**, 155–176 (2009)

Scheid, B., Quiligotti, S., Tran, B., Stone, H.A.: Lateral shaping and stability of a stretching viscous sheet. European Phys. J. B **68**(4), 487–494 (2009)

Scherer, G.W.: Editorial comments on a paper by gordon s. fulcher. J. American Ceramic Soc. **75**(5), 1060–1062 (1992)

Schultz, W.W., Davis, S.H.: One-dimensional liquid fibers. J. Rheology. **26**(4), 331–345 (1982)

Schultz, W.W., Davis, S.H.: Effects of boundary conditions on the stability of slender viscous fibers. J. Appl. Mech. **51**(1), 1–5 (1984)

Schultz, W.W., Zebib, A., Davis, S.H., Lee, Y.: Nonlinear stability of Newtonian fibres. J. Fluid Mech. **149**, 455–475 (1984)

Shah, Y.T., Pearson, J.R.A.: On the stability of nonisothermal fiber spinning. Ind. Eng. Chem. Fundam. **11**(2), 145–149 (1972)

Shah, Y.T., Pearson, J.R.A.: On the stability of nonisothermal fiber spinning-general case. Ind. Eng. Chem. Fundam. **11**(2), 150–153 (1972)

Shin, D.M., Lee, J.S., Kim, J.M., Jung, H.W., Hyun, J.C.: Transient and steady-state solutions of 2D viscoelastic nonisothermal simulation model of film casting process via finite element method. J. Rheology **51**(3), 393–407 (2007)

Silagy, D., Demay, Y., Agassant, J.-F.: Study of the stability of the film casting process. Polymer Eng. Sci. **36**(21), 2614–2625 (1996)

Silagy, D., Demay, Y., Agassant, J.-F.O.: Stationary and stability analysis of the film casting process. J. Non-Newtonian Fluid Mech. **79**(2), 563–583 (1998)

Smith, S., Stolle, D.: Nonisothermal two-dimensional film casting of a viscous polymer. Polymer Eng. Sci. **40**(8), 1870–1877 (2000)

Smith, S., Stolle, D.: Numerical simulation of film casting using an updated Lagrangian finite element algorithm. Polym. Eng. Sc. **43**, 1105–1122 (2003)

Smith, S., Stolle, D.F.E.: Draw resonance in film casting as a response problem using a material description of motion. J. Plast. Film Sheet. **16**(2), 95–107 (2000)

Strogatz, S.H.: Nonlinear Dynamics and Chaos: With Applications to Physics, Biology, Chemistry, and Engineering. Westview Press, Cambridge (2000)

Suman, B., Kumar, S.: Draw ratio enhancement in nonisothermal melt spinning. AIChE J. **55**(3), 581–593 (2009)

Taroni, M., Breward, C.J.W., Cummings, L.J., Griffiths, I.M.: Asymptotic solutions of glass temperature profiles during steady optical fibre drawing. J. Eng. Math. **80**(1), 1–20 (2013)

Van der Hout, R.: Draw resonance in isothermal fibre spinning of Newtonian and power-law fluids. European J. Appl. Math. **11**(02), 129–136 (2000)

van der Walt, C., Hulsen, M.A., Bogaerds, A.C.B., Meijer, H.E.H., Bulters, M.J.H.: Stability of fiber spinning under filament pull-out conditions. J. Non-Newtonian Fluid Mech. **175–176**, 25–37 (2012)

Weinberger, C.B., Cruz-Saenz, G.F., Donnelly, G.J.: Onset of draw resonance during isothermal melt spinning: a comparison between measurements and predictions. AIChE J. **22**(3), 441–448 (1976)

White, J.L., Yamane, H.: A collaborative study of the rheological properties and unstable melt spinning characteristics of linear and branched polyethylene terephthalates. Pure Appl. Chem. **57**(10), 1441–1452 (1985)

Willien, J.-L., Demay, Y., Agassant, J.-F.: Stretching stability analysis of a Newtonian fluid: application to polymer spinning and glass drawing. J. Theor. Appl. Mech. **7**, 719–739 (1988)

Yeow, Y.L.: On the stability of extending films: a model for the film casting process. J. Fluid Mech. **66**(03), 613–622 (1974)

Yun, J.H., Shin, D.M., Lee, J.S., Jung, H.W., Hyun, J.C.: Direct calculation of limit cycles of draw resonance and their stability in spinning process. Nihon Reoroji Gakkaishi **36**(3), 133–136 (2008)

Zavinska, O.: Spectral Analysis of Polymer Elongation Processes. PhD thesis, Technische Universiteit Eindhoven (2008)

Zavinska, O., Claracq, J., van Eijndhoven, S.: Non-isothermal film casting: Determination of draw resonance. J. Non-Newtonian Fluid Mech. **151**(1–3), 21–29 (2008)

The manufacturer's authorised representative in the EU is Springer Nature Customer Service Centre GmbH, Europaplatz 3, 69115 Heidelberg, Germany. If you have any concerns regarding our products, please contact ProductSafety@springernature.com

Printed and bound by CPI Group (UK) Ltd, Croydon, CR0 4YY
30/09/2026
02225500-0001